D0458818

AFTERMATH, INC.

AFTERMATH, INC.

CLEANING UP AFTER CSI GOES HOME

WITHDRAWN

Gil Reavill

GOTHAM
BOOKS

GOTHAM BOOKS
Published by Penguin Group (USA) Inc.
375 Hudson Street, New York, New York 10014, U.S.A.
Penguin Group (Canada), 90 Eglinton Avenue East, Suite 700, Toronto, Ontario M4P 2Y3, Canada (a division of
Pearson Penguin Canada Inc.); Penguin Books Ltd, 80 Strand, London WC2R 0RL, England; Penguin Ireland,
25 St Stephen's Green, Dublin 2, Ireland (a division of Penguin Books Ltd); Penguin Group (Australia), 250
Camberwell Road, Camberwell, Victoria 3124, Australia (a division of Pearson Australia Group Pty Ltd);
Penguin Books India Pvt Ltd, 11 Community Centre, Panchsheel Park, New Delhi–110 017, India; Penguin
Group (NZ), 67 Apollo Drive, Rosedale, North Shore 0745, New Zealand (a division of Pearson New Zealand
Ltd); Penguin Books (South Africa) (Pty) Ltd, 24 Sturdee Avenue, Rosebank, Johannesburg 2196, South Africa

Penguin Books Ltd, Registered Offices: 80 Strand, London WC2R 0RL, England

Published by Gotham Books, a division of Penguin Group (USA) Inc.

First printing, May 2007
10 9 8 7 6 5 4 3 2 1

Gotham Books and the skyscraper logo are trademarks of Penguin Group (USA) Inc.

LIBRARY OF CONGRESS CATALOGING-IN-PUBLICATION DATA
Reavill, Gil, 1953–
    Aftermath, Inc. : cleaning up after CSI goes home / Gil Reavill.
      p. cm.
    ISBN 978-1-592-40296-0 (hardcover)
    1. Crime scenes—Case studies. 2. Criminal investigation—Case studies. 3. Forensic sciences—Case
studies. I. Title.
    HV8073.R43 2007
    363.25'2—dc22

                                                                                              2006102542

Photo credits appear on page 285.

Printed in the United States of America
Set in Dante MT with Foundry Gridnik • Designed by Sabrina Bowers

For Eric Saks

socii criminis

Contents

AFTERMATH, INC.

You sending the Wolf? . . . Shit, Negro, that's all
you had to say!—Samuel L. Jackson as Jules in *Pulp Fiction*,
on the imminent arrival of the crime scene cleanup facilitator "the
Wolf" (Harvey Keitel)

The truth of things lies in the aftermath.—Sophocles

Author's Note

With a few exceptions (most notably the Dettlaff-Belter family), the names and identifying particulars of the crime victims in this book have been changed for reasons of privacy.

CHAPTER ONE

"You Could Smell the Hate"

Nicole Brown Simpson's front walkway

I write of the wish that comes true—for some reason, a terrifying concept.—James M. Cain

Mil Grabbers [are] designed to be used for entry and egress in a MOUT [Military Operations in Urban Terrain] environment....—Instruction manual, TRG grappling hook

After all his elaborate preparations for dealing out death, Nicholas Mazilli wound up knocking on the wrong door.

On a sweltering night in August 1998, Mazilli kicked things off by setting fire to his own trailer home outside Joliet, Illinois. He carefully made a videotape record of the conflagration.

His co-workers at the cavernous metallurgy shop of Northwest Tool and Die in Chicago Heights had been needling him, relentlessly twisting the blade every single day the way herd-mentality males sometimes do.

"Hey, Nick," one had said to him the day before. "I think I smelled Pam's perfume on Tommy just now." Hoots of laughter rang out from the half-dozen machinists.

Pam Mazilli was Nick's estranged wife. Tommy was his co-worker and best friend of twenty-five years, Thomas Johnson. Pam and Tommy had taken up together, and the thought of it was eating Nick Mazilli alive.

Maybe it was the heat, the succession of suffocatingly humid ninety-degree dog days. Maybe it was Nick having to endure seeing Tommy's shit-eating grin at work every day. Maybe it was the boiling tension he'd felt since he served as a Marine in the First Gulf War. Whatever it was, Mazilli snapped. After he burned down the green-and-white Kenilworth trailer in which he had been living ever since he and Pam split up, Mazilli headed east on Illinois Route 30, passing through the dark, heat-thick countryside across the state line into Indiana. He videotaped the drive too.

As the tape rolled, catching glints of all-night convenience stores and the occasional streetlamp, Mazilli's face was lit by the green glow of the dashboard light. "Tonight's the night," he said in a tight, choked voice, turning his face from the road to look into the lens of the video camera. "It's happening right now."

Cradled in his lap was a Remington 12-gauge pump-action shotgun. An AR-15 assault rifle lay beside him on the front seat with five hundred rounds of steel-jacketed ammunition, along with a Beretta 9mm machine pistol. He wore dark clothing and had carefully camo'd his face with charcoal. In the front pocket of his jacket he carried a can of lighter fluid and a butane lighter.

The videotape record was for his mom. "I want you to know, Mom, why I'm doing what I'm doing," he said into the camera. "I want you to tell them why this happened."

He continued, his voice robotic: "I've been going to bed every night wishing that Tommy was dead, and then waking up the next morning wishing he was dead. So that's what I'm going to do."

He arrived at the apartment building where his estranged wife lived with Tommy, in Merrillville, ten miles south of Gary. Mazilli, in full commando mode now, used a TRG military "Grabber" grappling hook and a nylon rope to scale forty feet up past the security gates to an open-air vestibule on the third floor.

But after all his preparation, he didn't know which door led to his wife's apartment. He chose wrong, knocking on the door of Apartment 3B.

Rachel Peterson answered with her baby daughter, Molly, in her arms. Her toddler son sat at the kitchen table, eating cereal.

She stared wide-eyed at the heavily armed ninja commando standing before her.

"Where's Tommy?" Mazilli demanded. "Does Tommy Johnson live here?"

Wordlessly a frightened Peterson gestured across the hallway to Apartment 3A.

Mazilli moved toward 3A, then turned back to Peterson. *This is it,* the paralyzed woman thought, *I'm dead.* She started to curve her arms protectively around her baby, ready to plead for her life.

"Get in the bathtub," Mazilli ordered her, employing the gruff command voice he had picked up in the Marines. "Get in the bathtub and pull a mattress over yourself and your kids."

The woman didn't move.

"Do it now!"

Rachel Peterson slammed the door shut and did as she was told. But she brought a phone into the bathroom with her and dialed 911.

Nick Mazilli planned on beating any police response. He didn't waste time. He strode across the hall and used his machine pistol to blast a pattern around the door lock of Apartment 3A. Nine 9mm bullets tore huge chunks out of the wooden door. The noise was terrific, but the world was muffled now by his orange foam-rubber earplugs. By the time he was finished, the lock mechanism was attached to the door only by splinters.

Mazilli kicked open the door and bulled his way inside, the Remington 12-gauge at ready position.

The door-blasting had taken only seconds, but Tommy Johnson was ready for him. He knew Nick well, knew his penchant for weaponry, and in his worst fantasies knew that something like this might happen. Johnson had armed himself with a 9mm Colt automatic of his own. He had just come out of the bedroom hall to the right of the living room area when Mazilli barreled into the apartment.

Johnson raised the Colt and fired, hitting Mazilli squarely in the chest.

It was a kill shot, and the night would have ended there, fodder perhaps for an NRA newsletter on firearm owners protecting their homes. But Nick Mazilli wore Kevlar body armor that night. Johnson's bullet slammed into his torso with the impact of a baseball bat home-run swing, but the projectile didn't penetrate and the blow didn't stop him.

Mazilli racked the Remington and pulled the trigger. A tightly choked pattern of triple-zero buckshot pellets hit Johnson's left thigh, dropping him to the floor of the living room. Mazilli stalked up to the fallen man and kicked the Colt handgun away.

"Pam!" Nick bellowed. "Pam!"

Pam Mazilli cowered in the bedroom, one room away.

"No, Nicky!" she screamed, sobbing hysterically. "Don't do it, Nick!" She curled up in a ball and clutched a photograph of the couple's three-year-old son, Aquino, holding it up toward the door to the living room like a charm to ward off evil.

"Motherfucker," Nick Mazilli said, looking down at his former best friend. "Cocksucker."

He switched weapons, bringing up the AR-15 that he had been carrying slung across his back. Firing methodically downward, he blasted at Tommy Johnson's left arm, severing it completely from his body. Then he adjusted his point-blank aim to do the same to Johnson's right arm, and both his legs.

The fusillade of bullets tore through Johnson's body, ripping apart the carpet beneath him and blowing cavities into the concrete underfloor below. Blood, bits of flesh, and bone fragments exploded everywhere. The supersonic assault-rifle rounds created pressure pockets as they sped through the air, and blood and body fluids were sucked into these pockets, so that the gore splashed back at Nick Mazilli as he fired and splattered across the walls and ceiling beyond him too.

Mazilli stopped firing. Police would later estimate that he expended 350 rounds of ammo tearing Tommy Johnson's body apart.

"Just kill me, Nick," Johnson begged, miraculously still alive but groaning in pain. "Just kill me." Mazilli could hear Pam's terrified sobs from the bedroom.

He was not done yet. He poured lighter fluid over Johnson's crotch area and set his testicles on fire. After allowing the flames to burn for a long beat—his best friend screaming at his feet, his estranged wife weeping one room away—Mazilli used his machine pistol to put a coup de grace bullet through Tommy Johnson's heart.

. . . .

I first heard about Tommy Johnson's murder from Tim Reifsteck and Chris Wilson, who own and operate Aftermath, a bioremediation company they founded in 1996 to clean up crime scenes after police forensics specialists had gone home.

Because I had spent a lot of time writing about crime, I was naturally intrigued by the level of violence, knowing the kind of overkill that Nick Mazilli exhibited was a hallmark of crimes of passion. The broken heart doesn't just want to kill. It wants to obliterate.

Later on, I would talk to some of the principals of the crime, those who had survived, and to the detectives and police who were involved. I'd get a fuller picture than the thumbnail Reifsteck and Wilson had given me. All standard operating procedure for a crime writer. But talking to Tim and Chris allowed me to think about a question that, oddly enough, had not occurred to me over a full two decades of writing about the violence humans do to each other.

What happens afterward?

Reifsteck and Wilson were at a funeral when their crews arrived to deal with the mess in Apartment 3A.

"Our cell phones started going off—boom, boom, boom," Reifsteck recalled. "We could tell the calls were all from our crew. We kept clicking them off, but as soon as we did, another one would come in."

"When I finally answered one of the calls," Wilson said, "I was pissed. 'What? You know we're at a funeral.'"

The crew chief sounded panicked, though, like a young child left alone at home. *You've got to come out here right away. We don't know what to do.*

At the same time that Reifsteck and Wilson's Aftermath crew recoiled at the carnage in Apartment 3A, Pam Mazilli's cohorts at the insurance office where she worked were receiving a bouquet of flowers that had been sent by Nick Mazilli. The note with the flowers extended Nick's condolences upon Pam's death.

But Pam Mazilli hadn't died the previous evening. Because she was the mother of Nick's son, Nick decided to allow her to live. Still weeping, she slipped past her doomed husband and her dead lover, through the blood-spattered living room, and out the door.

"I didn't look at him, but I knew he was standing there at the window," Pam said.

Mazilli held police at bay for two hours before putting a bullet through his own head.

The crime was so horrific, the details so gory and extreme, that on the day after the murders the Merrillville City Council debated whether to have the whole apartment building torn down as the only way to expunge the violence of the crime.

Still in their funeral clothes, Reifsteck and Wilson arrived at Tommy Johnson's apartment on the heels of the crime scene technicians and the coroner's office. From a funeral to a slaughter-house.

"You could smell the hate in that room," Reifsteck says of Apartment 3A's living room. "The strange thing about it was that the bedroom next door was totally undisturbed. It was decorated all in white. So you had this perfect-looking room with white curtains, white carpet, and white bedspread, and then you took two steps into a room where every square inch was covered with blood and body parts."

If their crew was stymied by the abattoir atmosphere, Reif-steck and Wilson were not. They rolled up their sleeves and set to work. What was left of Tommy Johnson's body had been scooped up and taken to the morgue by the coroner, but where he had died exhibited a clear silhouette, like a parody of a chalk body-outline. The blast marks of AR-15 bullets traced the dead man's torso. The burn mark at the crotch was still visible.

And everywhere, all over every surface in the twelve-by-fifteen living room, there was gore. The average human body contains about three six-packs' worth of blood. All of it seemed to be splattered at the scene. Victim's blood had mingled with killer's in a last ironic nod to poisoned friendship.

"It was like someone had taken a spray gun and painted the living room red," Wilson says. Pieces of brains, intestines, and bones remained embedded in the walls.

Reifsteck and Wilson quickly established a "clean zone" by

taping three-mil plastic sheeting over a twelve-square-foot section of floor near the front door of the apartment. They lay the
same plastic sheeting in long rolls down the stairwells on the way
to the ground floor, and out on the sidewalk to where their company trucks were parked in the apartment building cul-de-sac.
Their company logo was emblazoned across the side of red
GMC seventeen-foot trucks in twelve-inch-high orange letters.
Inside the trucks were plastic-lined boxes with red-and-black
"biohazard" labels on them.

Reifsteck and Wilson and every member of their six-man
crew wore white hazmat (short for "hazardous materials") containment suits and eye goggles, protection against blood-borne
pathogens that for all anyone knew were lurking in the remains
of Johnson or Mazilli. They set up two banks of thousand-watt
floodlights to illuminate the scene. They ripped up the blood-
soaked carpet, wrapped it, and carried it down to discard it in the
biohazard boxes.

Meticulously, they began to wash the room, using CaviCide
(a disinfectant and decontaminant) and hand rags, transferring
bloody and contaminated materials to the clean zone, then
bringing them downstairs to the boxes in the truck. Slowly they
expanded the clean zone. They scrubbed the loosely coagulated
blood that had collected in the small, cuplike holes blasted into
the concrete underfloor by Nick Mazilli's assault rifle.

The job took three days, over ninety man hours.

It was a situation you would never think about unless and
until it happened to you. Who cleaned up after tragedy? After the
millions of words spilled into print about the O. J. Simpson murders, no one ever bothered to write about who hosed down
Nicole Brown Simpson's breezeway (it was a team from the L.A.

sheriff's office). Or, for that matter, who cleaned up Sharon Tate's Benedict Canyon house? Or Ed Gein's horror homestead?

"The voice of your brother's blood is crying to me from the ground." So Genesis quotes God, after the world's first murder. Who cleaned up after Cain?

The Merrillville City Council was still debating whether to tear the building down. Neighbors from apartment buildings across the street dragged lawn chairs out to the curb and sat with beer coolers beside them to watch the cleanup. At times these onlookers broke into a spontaneous cheer when a particularly bloody bundle was carted out to the biohazard containers by the Aftermath crew.

That was okay. It was just another step of the long, strange trip that Tim Reifsteck and Chris Wilson had embarked upon when they formed their company. Aftermath was fast becoming the leader in the new field of "bioremediation," which involved taking care of the possibly contagious messes left behind when human beings shuffled off this mortal coil, in circumstances violent or otherwise.

Reifsteck and Wilson had learned the hard way to do their job well. No buildings needed to be torn down because a crime scene was too gory to clean. Three weeks after they finished with the apartment where Nick Mazilli had gunned down Tommy Johnson and mutilated his body, the place was rented to a state office worker and his wife, commuters who worked in Chicago. Because of the apartment's gory history, they got a break on the rent.

Except in photographs, I never saw Johnson's apartment. But when I began working for Tim Reifsteck and Chris Wilson at Aftermath, I encountered many scenes just as bloody, and a few that were bloodier.

CHAPTER TWO

Wisconsin Death Drive

The Mercedes G55 AMG "morgue mobile"

He du the police in different voices.

—Charles Dickens, *Our Mutual Friend*

Lady, the whole world is full of trouble.

—Steve McQueen, *Hell Is for Heroes*

Until I met up with the gents from Aftermath, I was a crime writer who had never actually been to a fresh crime scene. I wrote primarily for magazines, and for one magazine in particular, an oft reviled but absurdly popular "lad" magazine called *Maxim*.

Here's how one of my true crime stories for *Maxim* kicked off:

> By the time Roberto "Kiko" Rodriguez drove past his partner Johnny Boy's house, the yellow crime scene ribbon already fluttered around the yard. A police photographer's flash went off as he passed and Kiko saw Johnny Boy sprawled dead on the front lawn, a dark scarlet stain all over his head and back.

That kind of stuff.

All my stories for *Maxim* had the same theme: Young men (i.e., *Maxim* readers) behaving badly. Dealing dope, committing felonies, creating mayhem. Most prominently, dying. Publisher Felix Dennis labeled the kind of *Maxim* story I did "gritty reads." There was always one (and only one) per issue. They were, in fact, the only straight-faced articles in the magazine.

The main thing to realize here is that I never actually *saw*
Johnny Boy lying in his yard with the back of his head blown off.
Bertrand Russell talks about the crucial difference between knowledge by description and knowledge by acquaintance. I got all mine
by description. I spoke to homicide detectives in a suburb of Detroit who described the bloody tableau to me, and I read accounts
in the *Free Press* and *Detroit News*. I also interviewed Kiko Ro-
driguez in an Alabama jail, and he told me about what he'd seen
too. The fluttering of the crime scene ribbon and the darkness of
the bloodstain I got not from the detectives but from Kiko, Cuban
coke distributors usually being more poetic than cops.

If I thought about it at all, I didn't think my crime scene virginity really mattered. Kafka never visited America to write
Amerika. The Beach Boys didn't surf. Jean Racine wrote about the
sea all his life without ever actually setting his eyes on it. Imagination. That was the stuff writing was made of. I could sit in court-
room libraries sifting through trial transcripts, and as long as what
I came up with was vivid and packed with bite-in-the-ass details,
the secondhand nature of my experience mattered not a whit.
Forget knowledge by acquaintance.

Then I encountered Tim Reifsteck and Chris Wilson and, in
an unguarded moment, asked them if they would permit me to
job-shadow their crews. As soon as they said they would, I began
to have second thoughts.

"What if I puke my guts out?" I asked my wife.

A squeamish crime writer. It sounded like a joke, but there it
was. Among my family I was known for embarrassing episodes
of queasiness, including an occasion when I upchucked on a
county fair kiddie ride (revolving teacups) on which I was accompanying my then-toddler daughter.

In the 1980s a twelve-year-old girl, Pendharkar Chandana of

Andhra Pradesh, India, retched and vomited for twenty-eight straight days, eventually bringing up pieces of her own stomach. She was later found to have a tumor pressing against the floor of her brain's fourth ventricle, the binding site where vomiting is triggered.

"There are pills for nausea," my wife said, the soul of reason.

I nodded. "Cannabis," I said. She rolled her eyes.

"Rolling your eyes after your partner makes a statement is a marker for divorce," I said. She rolled her eyes again.

My wife is my moral compass. In fact, I sometimes think of her that uppercase way, as the Moral Compass. Since I haven't got much of one myself, she is handy to have around.

Of course, she wasn't the one headed, not to put too dramatic a point on it, into the jaws of death. The Moral Compass was a writer also. But she didn't write about crime. She found the whole subject matter of Aftermath pretty unsavory. But she was perfectly happy to see me go out and scrape brain matter off walls in service of our mortgage.

"It's a great idea, and I think you'd have a good time doing it," the Moral Compass said.

A good time? She didn't mean it that way. She meant that any crime writer worth his or her salt should, by decree of the crime writers' guild, jump at the chance to experience crime scenes firsthand. I had no choice. I either had to move forward or forfeit membership in the guild.

We were about to meet Tim Reifsteck and Chris Wilson for the first time that afternoon at our home in suburban Westchester. My wife informed me that she felt a shade uneasy about shaking hands with the owners of Aftermath.

"All I can think about is where those hands must have been," she said.

I spoke to Chris and Tim on the phone at their hotel in New York. I told them they could catch a train at Grand Central, and that I would pick them up at the station in our home village. We'd have lunch, talk a little, try to see if we could work together.

Something was wrong. I kept giving them Metro-North timetables, suggesting trains, trying to arrange for a schedule.

"We'll take a car up," Tim said.

"Well, the train is very simple," I said, trying to be helpful. "There's one that leaves Grand Central at one twenty-three, gets up here at two."

"That's okay," Tim said.

When they showed, I instantly realized that I hadn't quite understood what they were about. They were two Chicago guys on a trip to New York City, and they didn't want to ride a fucking commuter train. Their stretch limo was white and longer than my driveway. They made the driver wait while we talked.

....

Three months after I spoke to Chris and Tim that first time, I flew to Chicago's Midway Airport, rented a car, and drove west toward the suburb of Naperville. Naperville was next door to Plainfield, home of the corporate headquarters of Aftermath, Inc. I got lost on the way there and found myself on Ogden Avenue.

For crime enthusiasts (a weird, faintly moronic phrase, I realize, like "death enthusiast"), Chicago was hallowed ground. Ogden Avenue I knew as Al Capone's Main Stem, along which money, guns, alcohol, gamblers, and sexually enslaved women flowed back and forth from Cicero to Chicago. Ogden Avenue! It was like a Civil War aficionado suddenly stumbling across the Bloody Lane at Antietam.

I finally got my bearings and found my hotel, the Naperville franchise of the Extended Stay America chain. A boom in what the industry termed "long-stay lodging" had occurred in the last decade, with the number of rooms doubling. Average stay was forty-two days, so the hotels started offering home-away-from-home fringe services, like Easter egg hunts, cooking classes, and Crock-Pots in every room.

The Naperville Extended Stay was just one example of the trend. The desk clerks were trained to give the guests a big smile and a thumbs-up whenever they passed. It was supposed to make you feel more at home. When I checked in, the clerk asked me when my birthday was.

"Why?" I asked.

"We give a cake," he said. He was kidding I think.

The place should have been called Deracination Central. I could have been anywhere in America. The four-story hotel was located just off Illinois Route 59, behind yet another chain hotel, Red Roof Inn, and in front of a Steak 'n Shake fast-food outlet. I found my room, unpacked, and called Chris Wilson.

"I'm here," I said.

"We've got a three-week decomp up in a suburb of Milwaukee," Chris said. "Two of our technicians are going up there early tomorrow morning."

"A decomp?"

"A decomposed body," Chris said. "An unattended death. Eighty-four-year-old guy, heart attack, was lying there for three weeks before anyone found him."

I knew the Midwest had been in the grip of a horrible midsummer heat wave. A dead body three weeks in a closed house in August. Snakes on a plane.

"Fuck me," I muttered. "That's right, Chris, give the newbie writer from New York the worst, grossest, most filth-filled first job imaginable. Is this some sort of initiation rite?"

"I thought you came out here to go on jobs," Chris said.

"Homicides, massacres, four on the floor on Wonderland Avenue, six dead in the Nite Owl Coffee Shop, that sort of thing."

"Do you want to go or not?"

I spent an evil night. The hallway of the Extended Stay smelled nauseatingly of curry, and the stink seeped into the rooms. I tried to calm myself down.

Nothing human is foreign to me, I told myself, quoting the Latin playwright Terence. My erstwhile motto. Lots of people dealt with death and dying every day, and they managed to come through okay. Cops. The medical professions. EMTs. Morticians. It was actually not that rare a thing. Add them all up, and it was probably, what? Maybe a fifth of the population?*

In a fitful sleep that night, I was tormented by eidetic images of the leering, scabbed-over face of Regan Teresa MacNeil, Linda Blair's character in *The Exorcist.* "Your mother sucks cocks in hell," she hissed.

The next morning I dressed in what I thought might be appropriate trauma-scene clothing: a loose-fitting work shirt, jeans, and old sneakers. I drove thirty minutes through early-morning

---

* Using census figures and government vocational statistics, I eventually figured out a more exact estimate. Employing a very rough, nonstatistician's algorithm (law enforcement officers + "death care" workers + health professionals − chiropractors + chicken slaughterers, etc.), I determined that the sector of the populace who've become acquainted with the night was actually 8.5 percent, 25,262,190 people in the U.S. The percentages seemed to hold true internationally, at least in the West, with 5,043,522 working similar fields in the UK. The number employed by the hazardous-waste disposal industry in the U.S. was a tiny sliver of the whole, 8,580, or 0.002 percent of the general population.

traffic to the small industrial park where Aftermath, Inc., was headquartered.

Chicago. The City of the Big Shoulders turned a little flabby in its western suburban reaches. Here was the commonplace modern landscape: housing subdivisions, one after another, with more on the way. Every fifth vehicle on Route 59 was a cement mixer, heading out to pave over paradise. Naperville boomed. The city consistently topped lists of the best places to live in the country.

That morning it appeared hellish to me. Route 59 was a clogged commercial-strip artery, a ten-mile traffic jam. Discount superstores like Target, Wal-Mart, and Best Buy enthroned themselves along the road behind their vast, chessboard parking lots, with a row of chain restaurants ranked like pawns in front of them. Who ordained this specific arrangement? Was there some planning entity we could blame? Gaps amid the rampant housing developments still revealed the occasional cornfield, flat Great Plains farmland, already studded with placards announcing the zoning hearings that would allow it to be placed under the bull-dozer instead of the plow.

Tucked away down the small, horseshoe-shaped Arrowhead Industrial Park ("Amenity Movers," "Ken's Beverage," "Razzmatazz Lazy Daisy"), I found the two-story brick-and-sheet-metal ware house and office suite that housed Aftermath, Inc.

Living in New York, one of the few places in America (other than San Francisco, maybe, along with odd pockets such as Catalina or Fire Island) where the automobile is optional, I had fallen out of touch with how much car culture dominated this country. It was *real important* to people what they drove. They also wanted to know, and to comment upon, what you drove. I found Tim and Chris standing outside their headquarters that morning, admiring Chris's new purchase, a black Mercedes G55 AMG, an outlandish $92,000 vehicle that resembled nothing more than a morgue wagon.

"It sounds really cool, really loud," Wilson told me cheerfully.

Reifsteck nodded. "Pull up next to someone, they'd think it was a Harley beside them."

I'd seen strange, boxmobile SUVs on the road before: the Hummer, the Honda Element, the Scion XB. The G55 was the granddaddy of them all. We stood and stared at the top-of-the-line SUV some more.

"Wow," I managed. "That is a fantastic car." I meant fantastic in the sense of "like something unbelievable straight out of crazy film director Tim Burton's imagination," but Chris accepted it as "excellent," or "extraordinarily good." I visualized a child in an Edward Gorey drawing walking somberly in front of the thing, a black tulle scarf of mourning wrapped around his top hat.

"You ready to go?" Chris asked.

"Sure," I said. The image of *The Exorcist*'s demon-girl had vanished with the morning.

"Ryan and Dave are about to roll," Tim said, meaning Ryan O'Shea and Dave Creager, the two techs (Aftermath calls its crew members "technicians") assigned to the Milwaukee cleanup.

"You want to ride up with them or follow in your own car?" Chris asked. "What are you driving, a Buick?"

"It's what they gave me," I said lamely. I was afraid he was going to ask what size engine the rental had in it.

"It ought to get you there," Chris said.

"I guess I'll follow them," I said. I still held halfheartedly to the idea that on the way to Milwaukee, I could always bail out entirely, abandon the project and fly back home.

Ryan and Dave rolled around the corner of the Aftermath building in a white box-truck with a blue 2003 GMC cab. No identifying markings on the side. Aftermath used to trumpet its logos and services on its trucks, until Chris and Tim realized relatives of the deceased didn't always want "specialists in crime scene and tragedy cleanup" advertising on the street outside their homes. So, no more twelve-inch DayGlo orange lettering.

Dave and Ryan, the techs, didn't get out to greet me. Dave was driving, and we gave each other a wordless chin-rise greeting. Then they took off. I jumped in my rental, shouted good-byes over my shoulder to Chris and Tim, and tried to catch up.

They drove like bats out of hell. "This business is all about response time," Tim Reifsteck told me later. "Once a scene gets released by the police, once people realize what they've got in their house, they want it cleaned up *now*. We can get a crew to a site anywhere in the country in twenty-four hours or less."

"Not Alaska or Hawaii," Chris added.

"At least not yet" Tim said.

I followed Ryan and Dave as they hurtled east toward downtown Chicago on Interstate 88 (the Ronald Reagan Memorial Tollway), then headed north toward Milwaukee on I-294. They sped through electronic toll plazas, barely slowing down, and I tailed behind them, even though my rental wasn't equipped with an automated I-Pass, which meant it was an eighteen-dollar fine every time I did.

Repeating a jittery "normalcy" mantra to myself ("death is the most normal thing in the world," "no big deal," "X percent of the U.S. population deals with it every day," "melodramatic to think otherwise"), I kept my eyes pinned to the small orange "biohazard" sticker on the back of the Aftermath truck. I tried to calm myself down, but I couldn't.

We bounced north out of the maze of Chicago.

Wisconsin, I said to myself. *It had to be Wisconsin.* My native land. I was born in Wausau, in the dead center of the state. J. P. Donleavy has a line: "Under the sheep-gray skies of the land where I was born." Or Dylan:

> *My name it is nothing, my age it means less*
> *The country I come from is called the Midwest*

The country I come from. I did feel comfortable there, slipping into the Chicago landscape as though it were an old sweater. Where a lot of people saw flyover country I saw a familiar home. "A place of wide lawns and narrow minds" was what Hemingway called his native Chicago suburb of Oak Park, but I didn't find midwesterners narrow at all, in girth or sensibility. A genial, dogged, community-minded people.

In the present context, though, Wisconsin had turned into *the state in which I first saw a dead body.*

Chucky Sipple. A kid on my Little League team who I barely knew. When I was seven years old, he slipped from the piling of a Milwaukee Road railroad bridge that I could see from the bedroom window of my childhood home, fell into the ice-choked Wisconsin River, and drowned.

I remember the shock all us neighborhood kids felt, instinctively gathering the next morning at the death piling on the black-painted bridge, when Chucky's not-at-all-grief-stricken aunt steamed up. "I just want to see the place that dumb fuck went and drowned himself," she said.

You always remember your first dead body. The following afternoon at the funeral home, Chucky was a waxen figure arrayed in a coffin of polished mahogany, somehow more elegant in death than he had been in life, at least on the baseball diamond, where his fielding skills left something to be desired. W. C. Fields used to call death "the Fellow in the Bright Nightgown." For me, he was always a Little League shortstop.

"If I had known it was going to be an open casket," my coach apologized to parents afterward, "I wouldn't have brought the team."

Wisconsin. Home state of two of the most notorious murderers in American history, Ed Gein and Jeffrey Dahmer. Indicating that those who dismiss the place as boring and banal might be missing some darker currents that run beneath the cheesehead surface.

My real introduction to thinking about death came not via Chucky Sipple or Ed Gein, but from a cult text of the sixties underground, *Wisconsin Death Trip*, by Michael Lesy.

A spooky, chimerical, slippery book, still one of my favorites. I was disappointed to hear from one of my New York editor friends that although *Death Trip* was always a great critical darling, the book has never really sold that well, it was a loss leader, and the original publisher, Pantheon, kept it on its backlist simply for prestige value. The news somehow made me think less of my fellow humans, like the fact that Ike and Tina's "River Deep, Mountain High" never climbed past number forty on the pop charts. What's wrong with you people? Don't you recognize a classic when you encounter one?

For an adolescent boy growing up in deepest, darkest Dairyland, *Wisconsin Death Trip* sounded a thunderclap wake-up call. In the dusty archives of the State Historical Society of Wisconsin, Michael Lesy had uncovered an astonishing find: over two thousand glass photographic plates, their silver emulsion faithfully rendering life, and in the small-town Midwest, circa 1890, the work of a Black River Falls photographer named Charley Van Schaick.

Lesy matched Van Schaick's pictures of ordinary rural life with equally quotidian reportage from Frank and George Cooper, a father-and-son team of journalists writing for the *Badger State Banner,* which was a small-town newspaper in the same area that Van Schaick was a small-town portraitist. Lesy simply chose a selection of Van Schaick's photos and excerpted newspaper clips from the Coopers, but the effect was electric. It was only text plus image. But somehow one plus one made three, twelve, one million.

As Lesy catalogs his contents: "Ghost stories, epidemics, political careers, suicides, sales, insanities, bankruptcies, fatal abortions, medical testimonials and early deaths. . . ." There were numberless barn-burnings, deaths by ingestion of "Paris green" (an insecticide), and—most intriguing to me—an epidemic of destruction aimed at a newly developed novelty, the plate-glass window. The book presented a bygone world rocked back on its heels by tragedy, depression, and loss, a world of chaos that was light-years away from the bourgeois-fortress universe I inhabited as a teenager.

"Working men at Kenosha," reads one of Lesy's news clips, drawing a scene straight out of Aftermath's job files, "found the body of a man hanging from a rafter. The body was badly decomposed. Nothing was found to identify it."

In Michael Lesy's strange artistic calculus, life was portrayed as so inflected by death as to make the two an indivisible entity. To my early 1960s sensibility, accustomed to death being carefully pruned back, sanitized, and apologized for ("If I had known it was going to be an open casket . . ."), *Wisconsin Death Trip* was a book from which I would never recover.

Death-haunted. That's what Lesy's book managed to make Wisconsin. A pretty neat feat, given the thick-ankled Scandinavian-German vibe of the populace. The stolid landscape of my youth, where I had stood in cornfields during the summer and heard the eerie, rubbery squeals of the corn growing, turned out to have weight, majesty, meaning, after all. Like a southerner reading Faulkner. Oh, it says here that we are not just a species of dumb crackers. We are demigods worthy of a genius's attention. Those corn-growing sounds now reminded me of the high-pitched *"ree-ree-ree"* squeals from Hitchcock's *Psycho* shower-scene soundtrack.

I followed the bouncing orange biohazard sticker on a Wisconsin death trip of my own. We drove past Six Flags amusement park, the inverted-pretzel-shaped loop of its Bolliger & Mabillard Superman Ride of Steel roller coaster (a trip of "more than three minutes" at speeds "approaching 60 mph") visible from the highway. A pseudo pretend faux death trip for which you had to stand in line.

Vanity, saith the preacher. In my present death-haunted mood the whole concept of amusement seemed pure folly. How can those people laugh and scream when there is human decomposition occurring a mere thirty miles (a half hour via our excellent Interstate highway system) to the north?

I tried not to take myself too seriously, but I couldn't help it. The whole enterprise seemed impossibly fraught, and I gave myself over to unbridled freaked-outness.

We left I-90 at the suburb of Cudahy, on the southern edge of Milwaukee, along the lake, and drove east on Layton Avenue past General Mitchell International Airport, before plunging into a residential neighborhood. Our destination turned out to be a leafy street lined with houses of the preranch, post-Victorian, all-nondescript period of architecture. The residents, by the looks of their down-at-the-heel yards, clung to the lower ranks of the American middle class. A lot of them liked beige paint.

The quietest, calmest, most ordinary block in the world. Mickey Rooney and Judy Garland lived there, and Jimmy Stewart and Donna Reed. Summer vacation. Children, toddlers up to preteens, rumbled in small cowlike herds along the southern sidewalk. The northern side of the street was in shade, and deserted. "Keep on the sunny side," Mother Maybelle sang, "always on the sunny side." Visible to the east, through a four-block tunnel of maple and elm leaves, a freshwater-blue lozenge of Lake Michigan.

The orange biohazard sticker stopped bouncing in front of a two-story clapboard house, beige accented by white trim, with a redbrick foundation and light-green roof shingles. An enclosed porch, shuttered up tight. And, yes, on the sunny side of the street. Ryan and Dave got out of the truck. I parked the rental down the block in the shade and got out too.

"The sister or I guess the daughter or someone lives across the street," Ryan said. Balancing on the bumper, he rolled up the back door of the truck. "Dave's going to go over and have her do some paperwork."

I didn't know what to do. Ryan climbed into the interior of the truck. Enameled steel shelves, lockboxes, and tool drawers lined the front wall and ran down the driver's side. Plates of quarter-inch tread-quilted aluminum covered the floor.

"The first thing you do when you enter the truck—always, always, right away—is glove up," Ryan said. "We store a lot of biohazardous waste in here, and we keep it clean, but we have to play it safe."

Ryan O'Shea wore a navy-blue Aftermath T-shirt with white

lettering ("Specialist in blood cleanup—call 1-877-TRAGEDY"), size XXL to accommodate his build. He had indeed been All-State middle linebacker in high school in Plano, Illinois, a farming hamlet forty miles to the west of Naperville. He was twenty-three years old, six two, with big squared-off shoulders and a cheerful, open face. His dark blond hair was cropped close and wicked into short spikes, evidently the only acceptable male coiffure of the day, since Dave Creager and all the rest of the techs wore theirs the same way.

Creager was a half-hand shorter, bullet-headed but handsome, with an athletic build. The two had been friends since school. Creager used to approach O'Shea's house on his dirt bike, popping a wheelie for the whole length of the rural road. In turn Ryan would drive his truck straight across the farm fields to wind up at Dave's door.

Between football and an eventful leisure-time pursuit of off-roading, O'Shea had managed to break 18 of his body's 206 bones.

"I've broken my collarbone twice, once on each side, eight of my fingers, bones in both my hands, my arm, my jaw, blown out my left knee three times and my right knee once."

Ryan lowered the back door three quarters of the way down as he stripped to gym shorts. He instructed me to climb into a white level-one hazmat suit made of disposable Tyvek. Following his lead, I tied the suit off at the waist with a strip of yellow-and-black "police line do not cross" tape, effectively turning myself into a crime scene. He also told me to slip clear plastic booties over the suit's own built-in gray-soled boots. I appreciated the redundancy.

"This way," Ryan said, putting his own booties on, and tying them off with crime-scene tape too, "when your feet get mucked up and you're coming in and out of the scene, you don't have to take off and dispose of your whole suit. You can just toss the booties and put on new ones."

Ryan could tell I was anxious. He described what lay ahead. "Basically, on this job, anything that's contaminated with body fluid, if they died on a recliner or a sofa, that has to be removed."

First, though, he would enter the place with a heat-powered fogger and coat the whole interior with Thermo-55 disinfectant-deodorant. "The Thermo-55 sticks to everything and it smells like cherries," he said. "It's got an insecticide in it that helps kill the flies too."

If the fogger didn't do the job, the truck carried a machine that represented another level of response. "If we run the fogger and the smell is still very apparent, then we use the ozone machine," Ryan said. He slapped his hand on the gleaming metallic unit, stored on one of the truck's upper shelves. "We'll set that up and let it run overnight when it's real bad. It sucks everything out of the air."

I realized something before we ever made it into the house. *They're not just janitors.* I didn't yet know quite what the Aftermath techs were. I couldn't exactly grasp what the job entailed, but I knew it was a long way from the janitorial. During my college days, I briefly worked as a janitor on a work-study gig at the Engineering School of the University of Colorado. I never had to strip off plastic booties that were too mucked up with decayed body fluids to wear more than a single time.

Ryan showed me how to fit myself with a pink-and-gray 3M respirator with a clear acrylic visor, pulling it down over my face and then tightening the rubber straps behind my head.

"You're going to need this," he said. "It's going to be bad in there."

He tapped on my face shield as if to wake me up. "Listen,

here's the first rule: You treat everything contaminated with bio-matter as if it was poison."

I nodded.

"We don't know who this guy was or how he lived his life. He could have AIDS or hepatitis C. Probably not, but it's better to be safe. Okay? You got that?"

I nodded again.

Everything is illuminated? No, everything is contaminated.

Dave came back and geared up. We jumped back down onto the sidewalk. Gloved, masked, hazmat-suited and -hooded, we were ominous figures, exotic harbingers of Chernobyl, say, or Marburg in the middle of a mundane suburban landscape. If I looked out my own window and saw us outside, I would start thinking about which family photos I'd take along for the coming evacuation.

My emotions oscillated between the portentous and the pretentious. Everything seemed *meaningful*. I looked down to gray-pebbled sidewalk, cracked and aged. The children who had been coursing down the block when we pulled up (now removed a wary distance away) had run riot with sidewalk chalk. In addition to flowers, puppies and stick figures, they had traced their own bodies in cherry and lime, so there were chalk body-outlines everywhere, like in a noir movie.

The Milwaukee morning started out hot. I slipped my respirator up off my face, and I smelled the first whiff of decay from the still-closed house. Just a flavor, a faint tang of things to come, sugary and dark. My gorge rose.

"Are you doing all right?" Ryan asked.

I nodded and pulled my respirator back down.

"I'm going in first," he said. "Check things out, see what all we are going to have to do."

"I want you to do something," Dave said. "When we open the door, poke your head in without your respirator, before we fog it with deodorizer."

"Just to get the full effect," Ryan said.

We crossed the gaily colored chalk-decorated sidewalk to the front door.

CHAPTER THREE

The Samaritans

Chris and Jim

Which now of these . . . thinkest thou, was neighbor unto him that fell? —Luke 10:36

Observe the opportunity. —Ecclesiasticus 4:20

In spring 1996 Tim Reifsteck and Chris Wilson were making a solid income from their newspaper circulation and sales company, sending out crews of high school students to peddle subscriptions door-to-door. They had both recently graduated with business degrees from regional colleges—Tim from North Central College in Naperville, Chris from Eastern Illinois in Charleston—but they had known each other a lot longer than that, since the second grade in Sterling, Illinois, a blue-collar steel town west of Chicago.

Running the newspaper sales company required fast-and-furious work from about three o'clock in the afternoon until eleven at night. That meant during the day there was a lot of time left over for golf—which was what the two partners were up to that dreary, overcast morning in late April. They met at Tim's Aurora apartment building, planning on fitting in a quick prenoon nine, but were distracted by a collection of police squad cars and EMS vehicles parked in front of a house across the street. Tim didn't know these particular neighbors, but he was curious as to what had happened.

They walked over and spoke to a police officer at the scene.

"A kid committed suicide," the cop said. Then he added, almost as an afterthought, "They can't find anybody to clean it up."

"Really?" Tim said. "Did you try the phone book?"

"We already looked—nobody wants to touch anything like this."

Aurora, Illinois, the home turf of *Wayne's World*. The cop was bored, talky. He looked toward the house, a nondescript ranch that contained, somewhere inside, the blood-soaked aftermath of a tragedy. "It's really kind of a shame, you know? I wouldn't like to have the family do it. So we're calling around for someone to help out."

Tim and Chris exchanged the kind of telepathic look that old friends employ to indicate what they're thinking.

"Well, we're going golfing right now," Chris said to the cop. "But if the family needs our help when we get back, we'd be willing to pitch in."

All during their nine holes at Settler's Hill Golf Course that morning, they were distracted.

"Don't you think the coroner would take care of something like that?" Chris asked. "Clean up the remains?"

Tim said, "The coroner, or I thought maybe the police."

"Or a funeral home."

"Can you imagine what that would be like?" Tim said. "Your son kills himself, blows his brains out with a deer rifle, the police come and take away the body, but you have to go in there and make everything right again?"

"Well, they probably found somebody," Chris said, hitting a chip shot at the seventh-hole green.

"Yeah, they probably found somebody," Tim repeated.

Even with the distraction, they did okay on the links. Chris shot a 42 through nine, and Tim a 43.

When they returned to Tim's street the squad car was still parked outside the house, and the cop told them that the family had failed to locate anyone who would clean up the mess.

Had the boys been serious about their offer?

Chris and Tim looked at each other again. Yes, they were serious.

Chris attended church more regularly than Tim, but they both had at least a theoretical allegiance to the concept of the Good Samaritan.

"The Good Samaritan," said William S. Burroughs, "has probably gotten more people in trouble than any other story in the Bible."

Despite such rancid sentiments, Tim and Chris genuinely wanted to help a family in need. But they were also motivated by curiosity.

"We were interested to see what it would look like," Chris recalled. "We had never seen it before. Everyone has a picture of what a crime scene looks like from the movies, but it turns out the movie scenes are really mild compared to the majority of what is out there."

"You picture that in your head," Tim said, "and you just want to kind of know, does it really look like that?"

Neither one of them had any kind of military or EMS experience. But they had hunted and gutted deer. They didn't think the blood would be a big problem.

There was something else motivating them, too, something that was so far unspoken—but such old friends didn't need to verbalize something to make it real. They were young but not that young, both pushing thirty, a few years out of school and wondering if selling newspaper subscriptions was all that there was to life. Neither of them was satisfied. They both had their ears cocked for the knock of opportunity—especially if it involved helping out other people.

They canceled work for the rest of the day and went into the house to examine the scene.

The family was absent, out making arrangements for the funeral. The cop directed Tim and Chris down the stairs to the basement den. It was by now late morning, six or seven hours after the teenage boy had placed the muzzle of a .30-06 deer rifle into his mouth and pulled the trigger.

The scene remained fresh. The blood was darkening, but still glistened, wet. The only light in the room was the dim washed-out kind spilling from a pair of window wells. Most of the blood-spatter was contained on the carpet. But the scene gave mute testimony to how the deed had been done, where the boy stood, where the impact of the bullet exploded the blood from his body.

The house was silent. The cop stayed upstairs. The wood-paneled basement den was like a snapshot of tragedy.

"I had a definite idea about what a crime scene looked like from TV," Chris recalled. "There's a little pile of blood here, another pile there, and that's it. Well, a rifle leaves a little more damage than that."

It is actually not a simple feat, to commit suicide by deer rifle. You have to hold the muzzle steady and at the same time reach down and push—not pull—the trigger. The unnatural maneuver is called a reverse squeeze in forensic parlance, and oftentimes it results in a jerk of the rifle and a missed shot. (Chris and Tim recall one of their jobs was a failed suicide by rifle, where a teenage boy who had only managed to wound himself complained to his father, "See, Dad, I can't do anything right.")

Chris and Tim trooped back upstairs to look under the kitchen sink for cleaning supplies. They had brought over a pair of rubber dishwashing gloves and a bottle of bleach from Tim's apartment. In the house they found sponges, a plastic bottle of Mr. Clean, and another pair of rubber gloves.

"You boys all set, then?" the cop asked.

"Sure," Tim said. "No problem."

"I'll come back and check on you." Then he left the house, got in his squad car, and drove away.

They went back downstairs. They were unclear how to begin. Using a utility knife, Chris cut out a section of blood-soaked carpet. Tim scrubbed down the paneling. The light was not good. They couldn't really see what they were doing.

But when the cop came back three hours later, he seemed to be satisfied. "I want to thank you guys," he said. "You just saved this family a heap of heartache."

He ushered them out of the house and locked the door behind them. Chris and Tim dragged the two garbage bags stuffed with bloody carpet remnants, rags, and paper towels to the curb. The cop gave a little farewell honk of his horn as he drove off.

"Nowadays we would have done that job so differently," Chris said, looking back. "We cut out the section of carpet; now all of it would go. The entire room would have been scrubbed down and biowashed, a three-step chemical process. We would have brought in lights and just flooded that room with light. If we had to, we would have brought in blue lights that show up blood."

"We were just scrubbing down the walls," Tim said. "There was probably all sorts of biomatter left in that room."

They never saw the boy's body. To this day neither Tim nor Chris has ever met the family of the victim. But that afternoon they came to a simple but crucial conclusion, one that altered the course of their future.

They realized they had the stomach for the work.

Tim and Chris knew little of what they were in for that first day, after they cleaned up the suicide across the street from Tim's apartment building. They may not have been experienced in the protocols of crime scene cleanup, but neither were they dumb. They realized that they had stumbled onto something.

A light went on. An "are you thinking what I'm thinking?" kind of light. It was Chris who articulated it first. "This could be a real niche business," he said, as the two of them walked back across the street to Tim's apartment.

For the next two weeks Tim and Chris were on fire, on the telephone constantly, calling coroners, police, funeral homes, real estate management companies. They always asked the same questions of anyone who would talk to them. Would this be a good business to start? Is there a need for this kind of service?

The answers always came back the same. Yes, yes, yes.

"I could have really used you just last week," a contact at the De Kalb, Illinois, coroner's office told them. "We had a terrible crime scene, and the family had church members volunteer to come in, but a few of them just lost it and wound up sick in the hospital."

It would turn out later that Chris and Tim did not ask some questions that they should have, but for now, that yes repeated over and over was all they needed.

Tim and Chris were pioneers in uncharted wilderness. There were no guidebooks on the subject, no industry standards since there wasn't any industry. They had stumbled upon a situation that resulted when a sea change overtook society without anyone quite realizing its full impact.

The sea change was the advent of AIDS and other blood-borne pathogens such as hepatitis C. Since the AIDS epidemic became widely recognized in the mid-1980s, the rules for dealing with biohazards—biological matter such as blood, body fluids, used hypodermic needles, flesh remnants—had shifted. Medical professionals were the first to recognize the shift. Police departments began developing anticontamination kits with gloves and other protective equipment. If someone had invested in latex futures around 1985 or so, he would have gotten in on the ground floor of a bull market.

Since the sea change, no one in traditional housecleaning or janitorial firms would touch anything involving blood or body fluids. It was just too dangerous, full of killer unknowns and deadly variables. Since the cleaning-service industry was not centralized (apart from a few large franchise firms such as ServiceMaster), no one registered that the rules of the cleanup business had been drastically transformed.

Meanwhile, human nature hadn't altered at all. People were still hacking up their relatives in murderous rages and blowing their own brains out. The blood was still spattering around in sunporches and bedrooms and taverns and nightclubs. Old folks died alone and their bodies rotted and dissolved into sticky, possibly contagious messes. All this was still happening, but no professional service was cleaning it up anymore.

Families of victims, those left behind in the wake of homicide, accidents, or suicide, were expected to cope, not only with grief and loss that was unbearable but also with the physical mess of violent death. Police and EMS teams were often unhelpful or downright dismissive. A crime scene cop would qualify as "caring" if he left the survivors with a can of ground coffee to sprinkle around the site, to soak up body fluids and (ineffectively) kill the stench. Sometimes, as in De Kalb, the cleanup would fall to woefully unprepared church groups.

Until Chris Wilson and Tim Reifsteck. Every successful pioneering business, from Microsoft, Xerox, and Ford on down, was formed from just this sort of intuitive grasp of fundamental change.

It took them a while to fully understand what they had. Business was excruciatingly slow. For the first six or seven months, they rarely had a job per month. They did not suffer, since they had their newspaper circulation business to fall back upon, but they were puzzled. They knew there were jobs out there, plenty of them, but they weren't getting many. What was going on?

The one question they had neglected to ask police officers, coroners, and morticians—all those folks who had said yes, yes, yes, we need your service—was a pretty basic one.

"Can you recommend us to the families of the deceased?"

It turned out the answer to that question, for a lot of complicated reasons, was no. Generally, a public entity such as a coroner's office or a police department could not legally recommend a for-profit service like Aftermath. Liability concerns were usually cited as the reason for this. But there was the old-boy network too. Chris and Tim were the new kids on the block, with a new service to boot.

After three months in business, they had done a number of paying cleanup jobs that they could have counted on the fingers of one hand. They might easily have concluded that they had been wrong about there being a market for their service. They could have folded up their tent and gone home. But they persevered.

In that first year, they did $37,000 worth of gross business, no pun intended. In the second year that total went up to $75,000. Every year, their gross doubled or tripled. When the amount of annual business came in at $250,000, and next year turned into

$500,000 and the next rose to $1 million, they realized their hunch about Aftermath had been right.

They slowly gained a level of expertise in a business they decided to make their own: cleaning up after crime. That's what they called their company at first: After Crime Clean-Up. That was before they really knew the business, before they realized that it entailed far more than crime scenes and would eventually embrace all of the new field of "bioremediation." Even though suicide is technically a crime (there's an old riddle: What crime is prosecuted only when it is not successful?), the families of those who kill themselves did not want a van with After Crime Clean-Up painted on it parked in front of their houses.

Chris and Tim gradually came to understand the precise service they were selling. A quadruple homicide in Harvey, Illinois, was a watershed, their first large-scale crime scene cleanup. Lucas Tavernier took a .357 Magnum—what's called a hand cannon on the street—to his girlfriend and his girlfriend's sister's family. Tavernier hanged himself in the basement and left only an infant alive upstairs.

When Chris and Tim arrived on the scene the day after it happened, they could trace the progress of the crime by the bullet holes. "The oldest daughter jumped up in the second bedroom, and Tavernier just turned that way and blew her away right there," Chris recalled. "You could see where the three holes went through her, because they went through the glass window right behind her too."

Chris and Tim were a bit shell-shocked themselves, and exchanged a look as they surveyed the scene. "It was like, 'What have we gotten ourselves into?'" Tim said.

The job was an early technical challenge because the sister and her husband had been killed while they lay in a waterbed, which burst and flooded the premises with a ghastly mixture of blood, water, and feces. Tim and Chris had learned by then to enter a scene wearing hazmat suits, and they would need the protection that day.

But because the children who had been killed were twelve and eight years old, the Harvey job drove home another sort of challenge in their new work—an emotional one.

"That was a wake-up call," Chris said. "That was a huge one. It was a nice house with an in-ground pool in the back. Then we walked in and saw walls and walls of pictures of the kids and the family." The scene made Chris want to run home and hug his daughter extra tight.

On the threshold of the house in Cudahy, I was about to get an emotional wake-up call of my own.

CHAPTER FOUR

The Human Stain

Who saw him die?

The worm's-eye view is so often the true one.
—Uncle Pio, in *Gilda*

There's flies in the kitchen, I can hear 'em
there buzzing.—John Prine

As the door opened, the stench swung out from the interior like a hammer, landing a blow to my nostrils, sinuses, and mouth, crushing them all, so powerful an odor it was as though my skin felt it and my ears heard the sound of it too.

I retched. When faced with danger the sea cucumber vomits up its entire digestive tract, afterward growing a new one. I hadn't eaten, and if we're hungry, our sense of smell is heightened. Unlucky me. So I retched. Again. And again. Nonproductive emesis.

"Are you okay?" Dave asked.

"Sure," I managed. Then I gagged again.

What was happening to me? I had never smelled anything remotely as intense before. It was like a drug rush, only in reverse. My mouth flooded with saliva, a predictable precursor to vomiting.

All odor is particulate. Meaning that with respiration the rotted amino acid particles from the deceased (free-floating volatiles as small as 0.00000000000007 of an ounce) had entered my nostrils and nasopharynx. These odorant molecules, known by such evocative names as cadaverine and putrescine, swept against mucosa cilia in my nose and stimulated specific olfactory receptor neurons. Nerve cells passed along the oooh-bad-smell news via axon action to my *regio olfactoria*, which dumped a quick shovel pass to neuroreceptors in the lateral medullary reticular formation in my brain. Long story short, I felt like hurling.

Another particulate by-product of human decay is gamma-hydroxybutyrate, or GHB, notorious in its pure form as a date-rape drug. Standing at the threshold of a stifling house in suburban Milwaukee, I was being date-raped by death.

From Mikey's (Elijah Wood) memorable reading of his English paper in the movie version of Rick Moody's *The Ice Storm*:

> Because of molecules we are connected to the outside world from our bodies. Like when you smell things, because when you smell a smell it's not really a smell, it's a part of the object that has come off of it—molecules. So when you smell something bad, it's like in a way you're eating it. This is why you should not really smell things, in the same way that you don't eat everything in the world around you—because as a smell, it gets inside of you. So the next time you go into the bathroom after someone else has been there, remember what kinds of molecules you are in fact eating.

The members of Mikey's English class react to this recitative with "embarrassed silence" (according to James Shamus's script for the movie), filing it, no doubt, under the category "facts I just really don't need to know, Mikey."

I turned my face away so that Ryan and Dave would not see me retch. They were evidently iron-stomached, but they reseated their respirators and indicated by hand gestures that I should do the same.

The deceased had expired in the hallway, just inside the front door. To the left and right I glimpsed rooms piled high with clutter.

Straight ahead was the kitchen. The kitchen windows had black-green shades pulled down on them, shielding out the bright summer sunlight. Then the shades moved. They weren't shades after all. Thick carpets of flies covered all the windows. Even through the respirator, the smell was terrific.

"The family put these down so they could walk through here," Dave said, referring to a layer of flattened cardboard boxes, discolored with a dark, oily liquid. "But it was probably not a good idea."

He raised the corrugated cardboard. It lifted with a sucking sound. Long, ropy strings of black gelatinous muck came up with it, then broke off and fell back into the mess congealed on the floorboards below.

The human stain. Philip Roth, in his novel, was talking about the metaphysical mark left by human life, or at least the socio-political mark, the taint that our species leaves on whatever it touches. But here was the human stain in its physical essence. An oblong sheen of dark biomatter, truly sickening to behold, a skid mark exudate on the knickers of life.

Ryan and Dave were all business. Dave propped the gooey cardboard against the wall of the hallway. Their job was to eliminate from the site any trace of biohazard, but first they had to ascertain what was contaminated and what was not. They would take a Sawzall to the contaminated floorboards. The stain and everything it touched would be extracted from the site, packed in the two-by-four "bioboxes." The fluid-soaked corrugated cardboard laid down by the family, for example, would itself be broken down and packed into boxes. Boxes inside boxes, Chinese style.

Ryan knelt alongside the stain, using a pry-bar to break off a piece of brown-varnished floor molding along the east wall, trying to determine the cadaver's "drip zone." He pried up a floorboard. "It's all into the subflooring," he announced to Dave, his voice muffled by his mask.

Sherwin B. Nuland, a doctor and author, wrote compelling "reflections on life's final chapter" in his best-selling book, *How We Die.* I was confronting what happens next. How We Decompose.

....

In life, our bodies exist in a constant state of war, a state of siege. The barbarians at the gates are microbial. A few of them are fifth columnists already within the gates: *Staphylococcus, Candida, Malasseria, Bacillus,* and *Streptococcus,* which all humans harbor in their intestinal tracts. The beasts within are normally kept in check by the body's police-state enforcers from the immune system, but death means no more cops, and anarchy.

The last dated mail in the mailbox of the tan clapboard house bore a postmark of July 1. Let's take noon of that day, a Friday, as our TOD—time of death.* To mark the moment, the deceased may have emitted a death rattle, the last respiration before expiration, caused by the loss of the normal cough reflex to clear the mucus in the throat. The last breath "rattles" across the clogged phlegm, causing the eerie sound.

Human decomposition began almost immediately, approximately four minutes after death, at 12:04 P.M. By around

---

* Actually, TOD is itself a convenient fiction, an inexact concept useful in crime investigation, say, or to satisfy the human predilection for absolutes. There is no such thing as the moment of death, a precise instant of expiration, a clear-cut demarcation between being and nothingness. Like most biological occurrences, mortality is a process. Even so-called instant death, from explosion or massive trauma, takes a period of time, however brief. Death is not a flipped switch. It is always a journey.

one o'clock, the body had cooled *(algor mortis)* to the ambient temperature, which that day in Milwaukee was seventy-five degrees. The heat wave had receded somewhat, but the actual indoor temperature within the stuffy, sunny-side-of-the-street house was probably a couple degrees higher. The body's cellular cytoplasm gelled *(rigor mortis)*, producing the well-known rigidity of the limbs—the stiffness from which the slang term *stiff* comes.

Over the course of the next twenty-four hours, until around noon on Saturday, July 2, the blood settled in the lower parts of the body *(livor mortis)*, and the rigor disappeared. As the deceased lay facedown in the overheated Milwaukee hallway, spinal and brain fluids began debouching from his nostrils, eye sockets, and ears, providing an accessible buffet for bugs.

The insect assault on the corpse began immediately at death. Flies smell death, alerted by aromatic compounds released in amounts so small they are measured in parts per million or billion. By now the public has been carefully educated by repeated viewings of *CSI* to know that insects arrive in predictable, chronological waves, with houseflies and blowflies (including my personal favorite, the green hairy maggot blowfly—could a species *have* a more gross name?) as the first pioneers. The worms that crawl in and out and play pinochle on your snout are more specifically fly larvae.

We must imagine a clock—or better yet, a stopwatch. Only, this stopwatch exhibits a strange face indeed. Instead of numbers, it bears insect species. A small symbol of a blowfly, say, appears instead of the numeral 1. At every point along the chronological path of the stopwatch's ticking, specific bugs mark the "PMI," the postmortem interval, the time that has elapsed since death. Knowing the approximate PMI, we can thus discover the TOD, time of death—the holy grail of homicide detectives, which can rule out some suspects, focus in on others, and help detectives clear the case.

The bugs used to determine PMI are mostly of the kind Linnaeus, who developed the species classification system, called life's vilest: soft-bodied maggots passing through their three instars of growth, scavenger beetles, tiny parasitic wasps.

When an investigator catalogs the presence of certain insects on a corpse, noting their size and stages of development, it is as though he or she is reading the time on a stopwatch. Adjusting for temperature and other environmental factors, the investigator determines the time marked on the insect clock, and can testify with relative (and approximate) assurance when the deceased ceased.

"Who saw him die?" asked the English folk song about Cock Robin, answering the question with: " 'I,' said the fly, 'With my little eye I saw him die.' "

Flies sometimes find you even before your heart stops beating, functioning like police Breathalyzers, only in this case as "deathalyzers." A few molecules of butyric acid, indole, acetone, phenol, methyl disulfide, and other decay chemicals can be enough to alert the flying hordes. In general, members of the animal kingdom detect death much more efficiently than humans. A Tasmanian devil can smell a carcass a mile away. Barge operators towing the wreckage of the World Trade Center away from lower Manhattan for processing at Fresh Kills landfill reported that the flocking of seagulls signaled the presence of human remains in a particular load. Utilizing commercially available cadaverine and putrescine essences, handlers train cadaver-sniffing dogs for "HRD," or human remains detection.

By midday on that Saturday in July, insect eggs, laid in the first minutes after death mostly around the natural openings of mouth, nose, eyes, anus, and genitals, had hatched their larvae.

Over the next few days, maggots roamed in wolf packs over and into the body, hauling bacteria in their wake, secreting enzymes that further broke down tissue, and also out-and-out tearing at softer tissue with their mouth hooks.

Autolysis, or self-digestion, also got under way immediately after death, a process in which the unchecked cellular enzymes chew through the walls of the cells. Starved of nourishing oxygen, the body eats itself. It ferments. This happens sooner in such fluid- and enzyme-rich environments as the liver and the brain. Visual cues for autolysis include some sagging or sloughing of large sections of exposed skin, with small skin blisters forming as well.

There are ten trillion cells in the human body, mystically enough the same as the number of stars in the known universe. During autolysis, the cells of the deceased burst open by the billions. In the heat of the Milwaukee summer, the corpse began to melt like the Wicked Witch of the West.

When the cells rupture, they leak rich nutrients, which around Independence Day or so helped kick off the next stage of decay, putrefaction. Sulfhemoglobin lent the familiar greenish cast to the skin. Gases such as hydrogen sulfide, carbon dioxide, methane, ammonia, sulfur dioxide, and hydrogen built up within the bowels and other parts of the body, products of anaerobic fermentation. Though these gases sometimes cause the skin to rupture, normally they gain exit through the anus, and the death rattle is followed by a series of whistling-teakettle death farts.

By Monday, July Fourth, putrefactive bacteria and anaerobes, including micrococci, coliforms, diphtheroids, and salmonella, have joined the microbes already in residence. Various fungi, amoebae, pseudomonads, flavobacteria, and gliding bacteria joined in. The gang was all there by the time celebratory fireworks began to go up over Lake Michigan that evening, four blocks to the east.

"With the exception of micro-organisms living in deep-sea vents," writes leading forensic scientist Arpad Vass, "every micro-organism known is involved in some aspect of the human decompositional cycle, from *Acetobacter* to *Zooglea*."

The smell of putrefaction brought an additional wave of insects, including another one of my favorites, the cheese skipper, somehow a suitable presence in a Wisconsin death. "Cheese skippers," writes Jessica Snyder Sachs in *Corpse*, "prefer their bodies, like their dairy products, slightly aged." When startled, cheese skippers leap a couple feet straight up into the air, hence their name.

That Independence Day week and the next, predator insects showed up to feed not only on the body but on the maggot masses. Hister, rove, carrion, hide, ham, and carcass beetles arrived. Mites, coffin flies, and moths crowded in. Parasitoid wasps laid their eggs inside the soft bodies of the maggots.

By Friday, July 15, two weeks after TOD, the bloated body collapsed. The flesh exhibited a characteristic creamy consistency. The exposed skin blackened. The body fluids had drained, but mold grew on the liquescent underparts of the corpse, where it came in contact with the floor. The next week represented a last frenzied feast for the maggots, as the soft tissue of the body was pretty much consumed, leaving only ligaments and dehydrated skin, too tough for the delicate mouth hooks of the fly larvae. The remnants were left to the more powerful jaws of the beetles.

At this stage the body was discovered, on Tuesday, July 19. Left uninterrupted, the decomposition process would have continued through dry decay, or diagenesis. Given the environmental conditions (overheated interior space, no exposure to moisture), what remained of the deceased would probably have mummified, leaving the proverbial skin and bones (or "leather and stain," as Elmore Leonard has it). And hair: Apart from its susceptibility to fire, human hair is almost indestructible, resistant to water, rot, and many types of acid.

Chris and Tim tell of an early job on Warwick Avenue in Chicago that demonstrated what happens when remains are left for an even longer period than three weeks. The Office of the Public Guardian contacted Aftermath with a horrific situation. An elderly man had died two and a half years before. His mentally ill daughter did not report his death, and did not move the body at all. She merely threw a sleeping bag over the corpse and continued to cash her father's social security checks. Over the years, she put one air freshener after another in the house to mask the smell.

By the time the Office of the Public Guardian showed up, there were "thousands upon thousands" of air fresheners stuck up all around the premises. The daughter had duct taped the windows of the room where the old man's body lay to keep the stench of rot from escaping.

The measure didn't work. "When you were outside on the street you could smell it," Chris Wilson recalled.

For two and a half years the old man's body had lain unattended. Its fluids had drained through the bed and onto the floor, eventually, as with the Milwaukee job, working their way through the floorboards to collect in the basement. The rest of the body had thoroughly desiccated, turning itself into a granular "brown earth" powder, a mix of pupal cases, insect excrement, and bone. The "dust to dust" phrase from the Book of Common Prayer was quite literally demonstrated before Chris's and Tim's eyes.

When they arrived to clean up the scene, they noticed a snow shovel propped in the corner of the room. The coroner, it turned out, had tried to use the shovel to scoop up the desiccated body.

....

The Milwaukee County Medical Examiner's office removed the body, what was left of it, from the tan clapboard house in Cudahy a week before the Aftermath crew arrived at the scene. Coroners usually operate on a simple rule of thumb: They take whatever is attached. In cases of shotgun suicides, say, this can leave quite a bit of the dearly departed behind. Techs have picked up skull fragments, fingernails, eyeballs, eyelashes, nostrils, teeth with the roots still attached. Even in unattended deaths such as this, some remains always remain after the departure of the coroner (or, more probably, the body collector employed by the coroner).

At first, I didn't venture into the house, but observed from my post just inside the door. From where I stood I could see three clocks, one in the hallway still going at 1:10 (the wrong time), and two more, one stopped at 11:51 and one at 7:30. Off to the left, the living room held a small TV, two toasters, and a typewriter, as well as an incredible amount of clutter. Propped against a wall was a sampler spelling out the "Now I lay me down to sleep" prayer. *If I die before I wake . . .*

In a stuffed-full, curtained room to the right of the front door, stacks of old-time silent film videotapes (*Peter Pan, Uncle Tom's Cabin, Waxworks, The Cabinet of Dr. Caligari*) teetered next to yellowed *Milwaukee Sentinel* newspapers jumbled against pillars of blank VHS tapes. Movie posters lined the walls (*Little Lord Fauntleroy, Atomic Movie Orgy* Sponsored by Schlitz). Another faux-embroidered sampler: "Jesus, Mary and Joseph I give you my heart and soul." I counted seven tape recorders amid the disorder. A book: *You and the Law*. An old home-movie screen and a Bogen brand professional tripod.

I didn't know the name of the deceased. I tried to guess a life from the evidence that surrounded me. A movie projectionist?

"I'm going to check downstairs," Ryan said, gesturing me to follow him.

I stepped carefully into the front hall and around the edges of the stain. We passed through the kitchen. I approached the bombinating, fly-thick windows. Insect carcasses littered the kitchen sink, with tiny wasps picking through and feeding on the remains. The curtain of flies blocked the exterior light, casting the kitchen into shifting, shadow-patterned gloom. This was myiasis, fly infestation. Provided their progeny all lived, two ordinary houseflies could produce five trillion (5,000,000,000,000) offspring in one season.

Trying to mimic the scientific dispassion of an entomologist, I identified the striking iridescence of green bottle flies, which predominated, along with other blowflies, including blue bottles and attic flies, as well as orange-eyed flesh-flies and common houseflies. The housefly hums in the middle octave, key of F, but the kitchen windows sounded a half-dozen tones, with the deeper thrum of the blowflies the loudest. A fly opera.

As I followed Ryan down a narrow basement stairway, I passed a door to the outside. I could see a plastic children's play-set ten feet away in the next yard. I felt queasy. I decided to employ a method I had learned from prime-time television: If you grin, you can't gag. Physically impossible. *CSI*'s Sara Sidle (Jorja Fox) once pointed that out. I get all my antiregurgitation strategies from *CSI*.

Ryan looked back at me. Behind my face mask, my grin must have resembled a horrible, grimacing rictus. "What the fuck is wrong with you?" he asked.

"I'm okay," I said. I stopped trying to grin.

Downstairs, more clutter, stacked to the ceiling with only small aisles left for perambulation. The body fluids had drained from the deceased onto the floorboards upstairs, through the floorboards into the subflooring, then through the subflooring to dribble down the floor joists and drop onto a chest of drawers in the basement. On top of the chest of drawers was a large card-board box with a red-lettered Lucas-Milhaupt label, after a local machine company ("Your complete source for metal joining products and services").

The three-foot-square box was filled with multicolored industrial-strength extension cords, orange, black, yellow, and blue. The cords were partially obscured by a roiling, boiling, yellow-beige "maggot mass" of insect larvae, mostly meaty blowfly young'uns in their third instar of growth. I caught the sharp, distinctive scent of maggot excrement, smelling over-whelmingly of ammonia.

Maggots are even more gregarious than humans. They like a crowd. One tactic criminal investigators employ to gauge time of death is to thrust a thermometer into the middle of a maggot mass. The resulting measurement indicates how long the mass has been in operation. A maggot mass is capable of chewing through forty pounds of soft tissue in a day.

Looking into the box after Ryan hauled it down, I vomited a thin, pathetic gruel inside my mask. I knew there were such phe-nomena as "vomit waves" that infect hospitals, airplanes, and binge-drinking parties. The regurgitation reflex was contagious, a product, evolutionary biologists say, of our primate past. One member of the group hurled from a bad slice of fruit, say, and the whole chimp family, Cheetah to Jane, upchucked too. Like mutual grooming, only of the gut. This communal response was also useful among young partygoers, clearing everyone's stomachs to make room for more alcohol, a practice called boot and rally in the States, and tactical chundering in the UK.

Ryan seemed to be immune to the contagion. No vomit wave was going to catch him in the basement of the beige clapboard ("Catch a wave," sang the Beach Boys, "and you're sitting on top of the world"). He soberly scoped out the extent of contamination, eyeballing it close up. The spinal fluid from the deceased had rectified into a glaze-brown varnish pooled on the floor.

"That's the dangerous stuff," Ryan said, pointing to the fluid. "That's what's most likely to have the hep C virus in it, or HIV, bad stuff like that."

Cerebrospinal fluid. Aka CSF. Aka, in medical Latin, *liquor cerebrospinalis*. The saline solution in which the brain floats.

I had to back myself into a trash alley for Ryan to pass me on the way back upstairs. I followed him out of the basement.

"It's all down there too," he told Dave. "Maggots. A whole shitload of them."

The two techs conferred at the other end of the hallway. A shred of my prefatory research occurred to me. "Did you know a maggot breathes through its asshole?" I asked.

Ryan and Dave stared over at me. I immediately got the idea they did not need Johnny Britannica spouting fun facts at precisely that moment. But it was true. Most maggot species utilize anal spiracles for respiration, which, considering that they are nature's Dumpster divers and thus constantly upended, makes a rude sort of sense.

"Ryan's going to fog the place with Thermo-55 first," Dave

said. I had forgotten what Thermo-55 was. Disinfectant? I hoped it was the most effective deodorizer known to man, with larvicidal properties, developed by some top-secret arm of the military industrial complex.

I stripped my plastic booties off and followed Dave outside. Fresh air was a relief, mitigated by the fact that I still smelled the candied stench of decay no matter how far I removed myself from the death house.

Sense memory dictated I would forever smell what had assaulted my nostrils that morning. Some of my olfactory receptors adopted themselves to the odiferous decay molecules that had been presented to them, so the receptors were effectively locked into place, ready to identify the same stench again later. The problem is, sometimes these custom-tailored receptors get falsely triggered by other molecules, and we wind up smelling something from our past, something that is no longer there, in a phenomenon that olfaction experts call phantosmia.

It's true what they say about death. "Once you smell it, you don't forget it."

Olfactory memory is more accurate and long-lasting than visual memory. Furthermore, it is seated in the limbic part of the brain, the same region where emotions are generated. So the odor I smelled that day, and the emotional upheaval generated by it, would be inextricably linked, and with me forever.

The block's street urchins still maintained their distance, cowed by the sight of Tyvek spooks in their midst, but an adult who identified herself as "Phyllis, down the street," approached Dave.

"I just took my examination for the county police academy," she said, attempting to establish her bona fides. "I was wondering, what do you do to get into a field like this? Was it bad in there?"

Until recently, Aftermath always had male techs. "We kept trying to hire women, but they never worked out," Tim said. But now the East Coast territory has an all-female crew.

I would have been eager to discuss the job with Phyllis, with anyone. Ever since we had rolled up to the job, I felt the urge to call the Moral Compass and unload on her. But Dave wasn't forthcoming. As Phyllis peppered him with questions he answered with noncommittal grunts or single-word replies, continuing gathering together bioboxes and supplies for the task ahead.

Dave told me later, "I could talk with people for an hour, two hours, and they would still have questions. If I did that I'd never get any work done. I used to be a lot more patient when I first started, but now I just blow them off."

Phyllis hung in there. She started talking about the deceased. "He had the hoarder's disease," she said. "It's a real illness, you know."

She said he was a recluse too. "He didn't like to answer the door. When he wasn't around last month, I just thought he was in the hospital again."

"I feel terrible for the family," she said, casting her eyes to the daughter's house, across the street three houses to the west. "They've got a world of trouble."

The chatty Phyllis finally retreated. A Good Humor truck tootled down the street toward us, its chimes sounding, to my ears right at that moment, maniacal and unstrung.

"I don't get it," I said to Dave. "The daughter lives on the same block, but the guy is lying there dead for three weeks before anyone finds him. How did that happen?"

"They're real good people," Dave said. "They'd probably talk to you."

"Unless, you know, they were estranged."

Unattended death. Maybe the saddest two-word phrase in the language.

I didn't stick around to find out. Reeling from the physical and emotional reaction to what I had encountered inside the beige clapboard, I bailed out of the job, out of Wisconsin, and out of my journalistic responsibilities, beating my way back to the Extended Stay to recover the stomach I had lost.

It took me a while to recuperate, but I eventually went back to the house in the Milwaukee suburbs ("As a dog returneth to his vomit," the Bible says, "so a fool returneth to his folly"), to see what it was like when the Aftermath job was done.

CHAPTER FIVE

Father Time

Alois Felix Dettlaff

We don't want to be bothered.—Langley Collyer,
famous pack rat and recluse, on why he and his brother, Homer,
shut themselves off from the world

**Tonight my ambition will be accomplished. I
have discovered the secret of life and death.**
—Title card, Edison's 1910 *Frankenstein*.

"Collectionism," or obsessive hoarding, is just one of a battery of behaviors that make up Diogenes syndrome, a little-understood affliction that causes some elderly people to neglect hygiene and live in reclusion. "Animal hoarding," familiar in the person of the venerable "cat lady," can be another aspect of it.

Social workers report stories of a mummified litter of kittens discovered beneath the layers of trash, of rats chewing through the oxygen lines of elderly victims. The syndrome hits across de mographic lines.

"I've gone into places where people were extremely wealthy and well-educated and they're hoarding the same things that desperately poor people are hoarding," said one San Diego sheriff's deputy. "What I find almost everywhere is plastic bags. They seem to save those by the jillions. And restaurant napkins."

Hoarders inevitably tend toward extremes, saving food, even though it may be rotten, foraging compulsively in trash bins, displaying signs of defensiveness and paranoia. Howard Hughes, who kept his own urine collected in bottles, demonstrates the extreme end of an extreme spectrum: syndrome sufferers who refuse to throw away their own bodily waste.

The syndrome gained its name from Diogenes of Sinope, the first Cynic, an acid-tongued misanthrope ("In a rich man's house there is no place to spit but his face") who turned his back on the Athens of the Golden Age to live out his life in a barrel. The incidence of Diogenes syndrome in the general population runs about one person out of two thousand, meaning there are around 150,000 in America, 29,000 more in the UK.

The most infamous instance of hoarding was the Collyer brothers, Homer and Langley, who in 1909 moved into a four-story Harlem mansion at 128th Street and Fifth Avenue in New York City. They lived without gas, electricity, or telephone ("There is no one I particularly care to talk to," Langley explained), with ten grand pianos in the house that the Columbia-educated Langley played for his blind brother.

A PALACE OF JUNK read the headline over three decades after the brothers moved in, when authorities pried open the barricaded doors of the Collyer mansion in 1942, finding it filled with "rolling hills of neck-deep rubble."

But there was poetry inside too. "When Homer first lost his sight," Langley said, "he used to see visions of beautiful buildings, always in red. He would describe them to me and I would try to paint them just as he directed. Someday, when Homer regains his sight, I will show the paintings to him."

Safely back inside the Extended Stay stockade, I mused on the Collyer brothers and thought about the cluttered Milwaukee death house I had just visited. What if there had been poetry there too? I realized that for a brief interlude I had effectively lost my mind. I didn't take any notes. I forgot about journalism. I completely abdicated my function as an observer, descending instead into puerile weeniehood.

The Moral Compass called from New York. "How are you doing?"

"I'm fine," I lied.

I told her about the job. She didn't want details.

"Do you want to talk to your daughter?"

"Sure," I said.

"You don't have to get too graphic," my wife warned me.

I spoke with my daughter. Her voice had a soothing, stabilizing effect on me.

"I miss you," she said.

"I miss you too, darling."

A beat of silence. "Dad," she said, "why can't you write about something normal?"

I spent a fretting weekend and called Dave Creager on his cell the next Monday. I had seen the "before," I told him. Now could I see the "after"?

"Sure, come on back up," he said. "We couldn't finish this job because we got called away to a handgun suicide in Minneapolis."

"You interrupted one job to go to another job?"

"Oh, yeah, it happens all the time," Dave said. Suicides and homicides took precedence over decomps. Fresh blood always trumped decay.

I slowly grasped the nature of work at Aftermath, Inc. Going into the project, I had imagined a chain of prim storefront franchises scattered across the country, something like U-Haul, say, or Burger King, only full of cleaning supplies instead of moving boxes or hamburgers. Crime Scenes R Us. It wasn't that way at all. I began to understand the fundamentally itinerant nature of the business. There were three or four central Aftermath depots and some far-flung offices, out of which tech crews in white box-trucks would hie forth like Ringwraiths.

A two-man crew such as Ryan and Dave's might spend three weeks on the road, gradually filling the back of their truck with contaminated material. They might drive ten hours to a site in Michigan, turn around and hit a job in southern Illinois, then head out to Iowa or Minnesota. At the end of a string of jobs, their truck stinkfully packed with reeking bioboxes, they'd limp back into Arrowhead Industrial Park to the little warehouse on South Mandel Road in Plainfield.

Not a life for the faint of heart.

I headed back up to Milwaukee on Son of Wisconsin Death Drive 2: The Sequel, a little more clearheaded this time. The Aftermath truck was parked ass-end in at the front door of the tan clapboard house.

Ryan poked his head out of the back. "We're pretty much done," he said. "Dave went over to get the family, have them do a walk-through."

"You went to another job since I last saw you two days ago?" I asked him.

He nodded. "Up in St. Paul. It was pretty bad. A suicide, and the guy had a couple Irish setters that tracked his blood all over the house. It was a real mess."

Ryan offered me his digital camera. Aftermath techs always exhaustively document their jobs for insurance purposes. As I scrolled through the scenes of carnage, Ryan gave me a blow by blow.

"The husband and wife had a big argument," Ryan said. "I guess they were pretty drunk."

The wife passed out on the couch, while the husband went down to the basement, "where he kept his guns," Ryan said. He climbed back upstairs carrying a .44 with a target barrel on it, and shot himself in the head at the top landing. The cartridges were Magnum-load hollow-points, "which basically cause your head to explode," Ryan said.

"The wife stayed in a alcoholic stupor on the couch," he added.

"She didn't wake up?"

"I guess it was a deep sleep," he said, shaking his head. "She didn't hear the shot, but the neighbors did, and they called the landlord, who opened the door and found her lying there, still asleep. The sheriff's deputy said she had some of her husband's brains spattered on her."

I flicked through the camera slide-show of the job. The dogs, Ryan said, had made a little feast of their master before the scene was discovered.

"What is all this white stuff down here?" I asked Ryan, looking at a photo of a blood pool in the camera. "It looks like paper or something."

"Before he shot himself," Ryan said, "the guy cut up all his credit cards and scattered the pieces at his feet."

A tableau of misery: confetti evidence of his debt, plus an alcoholic wife passed out on the couch.

Dave came back from across the street. "The wife was there when we were working, and she asked me, 'Will you wash my dogs?' I said, 'Hell, no, lady, take them to the fucking vet or somewhere.'"

"Only I bet you didn't say 'fucking,'" Ryan said.

"Fuck, yeah, I did," Dave said, but Ryan and I both knew he hadn't. I had seen Dave with the family on this job, and while at most other times he could be raucous and bawdy, with grieving relatives he was the soul of politeness and discretion.

Dave gestured back toward the front door of the house. "You want to go in, see what the aftermath of an Aftermath job is like?"

"Do I need to gear up?" I asked him.

He shook his head. "Come in and see."

....

The transformation awed me. The stink, what there was of it, was coming from the material packed in bioboxes in the back of the truck. The house itself smelled . . . well, not like a summer garden, since it was still piled high with the musty detritus of a man's life, but nice enough.

"We like to mix TR-32 cherry with TR-32 mint," Dave said. "We call it candy cane." TR-32 was Aftermath's high-end deodorizing liquid, applied by spray or used directly in solution. It came in different perfumes, including lemon, cherry, and mint. The air in the house did indeed have a faint candy-cane scent, a North Pole, Santa's workshop smell, with chemical undertones, as if the elves were enlisted from DuPont.

The whole atmosphere was light-years away from what I had first encountered. Ryan and Dave had cut out the flooring in the hallway, leaving an irregular, Joan Miró–shaped hole, through which I could see down into the rat's nest of a basement. Where the biomatter had dripped down onto the floor beams, they had cleaned and then sprayed the aptly named Kilz sealant. The paint showed up bright white against the dingy surroundings. Every bit of material contaminated by "bio," as Ryan and Dave referred to biomatter, had been packed into the truck.

The whole place had undergone a three-step biowash. That

meant Ryan and Dave treated all the contaminated surfaces first with Microban, an enzyme cleaner, then with a foaming disinfectant called Spray & Wipe, and finally with the deodorizer.

"Three stages to remove the biocontamination," Ryan said. "Kill it, pull it away from the wall, and deodorize it."

The clutter, though, was left as it was. Aftermath's mandate stopped at bio. At $250 an hour, they were too expensive to be used as a moving service. The ceiling-high collections of the dead man were the responsibility of the family. I gazed over at the jumble of junk on the front porch. An old-fashioned scythe rested atop the pile, its four-foot blade papered over with a banner that read "Harvesting Movie Memories."

"Most people's lives," noted that original Little Miss Sunshine, Tennessee Williams, in *Suddenly Last Summer*, "what are they but trails of debris, each day more debris, long, long trails of debris with nothing to clean it all up but, finally, death."

"What was this guy's name?" I asked Dave.

He consulted his job sheet. "The deceased was . . . Mr. Dettlaff. Al Dettlaff."

The daughter and son-in-law came over from across the street to sign off on the site.

"I want to say up front, thank you," the son-in-law told Ryan and Dave. "You did a great job."

They introduced themselves to me as Helen and Tom Belter, and they walked through the transformed house of Helen's father marveling, as though it were indeed Santa's workshop.

"I can't believe it," Helen said over and over. She was a sad-eyed lady who looked as if grief had shaken her like a rag doll.

I followed Dave and the couple back to the Belters' house across the street. As Dave completed the paperwork, I spoke to

Helen and Tom. I told them I was a writer working on a book about Aftermath, and asked if I could speak to them.

"I wonder if I could talk to you about your father, Helen."

"You went in there?" Helen asked, clearly upset with the idea. Her family's dirty laundry.

"I was part of the crew," I said.

"He was part of the crew," Tom repeated, soothing his wife.

"I want to be able to write about your dad as he was in life, not as a cleanup job," I said.

"Right," Tom said, on my side, getting enthused. He had already picked up on Ryan and Dave's lingo. "He's not just bio."

He's not just bio. Exactly right. Wiry, small-statured, filled with Midwestern goodwill, Tom Belter solicitously protected his more fragile wife. Apologizing for the clutter ("My father lived here before he moved across the street, and I don't think the place has ever recovered"), Helen invited me into her perfectly presentable home. We sat at the dining room table and spoke over the course of an afternoon.

....

Alois Felix Dettlaff was a sedens, a useful word applied to someone who has never lived anywhere else but his hometown. He spent his whole life in Cudahy. As a child of eight, he took over the basement of his father's drugstore at Packard and Layton, only a couple blocks to the west of where we were now talking, and put on magic shows for the neighborhood. But Al Dettlaff's path in life was determined when his father, Alois John Dettlaff, bought him an 8 mm movie projector. Now silent movies supplemented the magic shows.

"Five cents admission, plus the audience members got a two-cent candy bar and popcorn," Tom Belter said.

It was the beginning of a lifelong obsession. And Al Dettlaff was always an obsessive man, thorny and irascible. More than anything else, he loved the muffled, time-travel world of silent films, especially early horror movies. He became a collector, scouring archives for such classics as F. W. Murnau's original *Nosferatu*.

Dettlaff worked as a quality control technician at machine shops around the area, including In-Land Company and Lucas-Milhaupt, but his real life revolved around silent films.

"He hated the new movies," Helen said.

"He never related well to people," Tom said. "If you came over, it was always, 'Hey, you want to watch a movie?' If you didn't, if you weren't into what he was into, you were stupid."

To encompass his movie-collecting passion, Dettlaff formed AD Productions and developed a new persona for himself: Father Time. He began marching in local parades.

"He got his ex-wife to sew some sheets together," Tom recalled. "He had a tricorner hat, which was good for the Bicentennial, and he hung a big hourglass around his neck on a red-white-and-blue tie, with a sign that read 'Father Time.'"

Marching in Milwaukee Fourth of July or Memorial Day parades, Dettlaff always carried his scythe, the same one I had seen on his front porch.

"I guess he was more of a Grim Reaper kind of character," Tom said. "He smoked a pipe and chewed tobacco, so the red-white-and-blue tie got a little stained and grotty."

Dettlaff's obsession took deeper hold of him the older he got. His only son, born Alois Felix Dettlaff, Jr., changed his name to James and wanted nothing more to do with his dad. Dettlaff himself became increasingly brusque, irritable, happy only when watching silent films.

Helen and Tom had eight children and fifteen grandchildren. But his progeny left Al Dettlaff cold. "He didn't relate much to his grandkids," Tom said, "because they couldn't talk about silent film with him."

"Biggest mistake I ever made," Dettlaff's father, Alois John ("a gentle smiling man," according to Tom), once confided, "buying him that damn movie projector."

But in a single stroke Al Dettlaff justified his hobby, snatching a priceless gem from cinematic oblivion. Thomas Alva Edison's 1910 *Frankenstein* had been placed on a list of lost films by the Library of Congress. The earliest cinematic treatment of Mary Wollstonecraft Shelley's epochal novel (a text that some credit with inventing the modern age), Edison's *Frankenstein* was a holy grail of silent-horror-film enthusiasts.

Father Time got his hands on a copy. A 35 mm nitrate print, the only one in existence. The print's provenance was obscure, even within the Dettlaff-Belter family.

"Al didn't like to talk about it, but I think he bought it from an aunt who was a projectionist," Tom said. "In those days nitrate film stock lasted only so many plays—then it was discarded. But projectionists didn't always throw it away."

Now Al Dettlaff's obsession had an object worthy of it. "From then on," Tom said, "that *Frankenstein* was his baby."

Dettlaff jealously guarded his find. He resisted releasing it on VHS, at one time spending seven years fussing over a transfer before abandoning it. He always insisted on showing it personally at horror convention screenings.

Finally, near the end of his life, he committed *Frankenstein* to DVD. Edison's film is an extraordinary work, still terrifying almost

a century after the fact (terrifying and, of course, given the styliza-
tions of the period, hilarious too). The monster (Charles Ogle, the
screen's first Frankenstein's monster) is created not from stitched-
together body parts, but conjured chemically, in a sealed cabinet.
In an impossibly eerie scene, a skeletal scarecrow frame grows
flesh and features by accretion, until the monster is revealed.

Watching the flickering, hundred-year-old images, I immediately recognized the process. It was human decomposition in reverse. What happened to Al Dettlaff's body in his stifling front
hallway that July was mirrored uncannily by his prized posses-
sion, Edison's silent film.

In the last years of his life, Dettlaff's body, beset by diabetes
and heart disease, was failing him. As he recovered from a bout
of illness at the veterans hospital, Milwaukee County social service authorities condemned his house as uninhabitable.

"We worked like dogs to get the house back in shape," Tom
Belter recalled. "He wanted to die in that house, not in a hospital
or in the VA domiciliary."

Together, Tom and Helen managed to make the place livable,
and their work passed inspection with the health authorities. But
the homecoming did not go smoothly. Dettlaff flew into a rage
when he saw how his home had been neatened and his possessions rearranged.

"It was supposed to be a celebration, but it wasn't," Helen
Belter said.

The light of an August evening slanted into the Belter dining
room. The smell of neighboring barbecues drifted into the
house. Helen became teary-eyed, and Tom put his arm around
his wife's shoulders.

"There was love in my dad too," Helen said. "We had a lot of
good times."

I tried to phrase the question delicately. "Helen," I said, "what happened at the end? Why didn't you visit him?"

I saw her wince. It was a knife that had entered her heart many times over the past few weeks.

"Helen did everything for him," Tom said. "When he was in the hospital, she had to constantly go back and forth getting things for him from the house."

"He would write down lists of things he wanted," Helen said. "First of all, it was hard finding anything in there."

"She did for him," Tom said, nodding his head firmly.

But during family therapy sessions at the veterans hospital, counselors saw a woman bowed down by caring for her impossible-to-please father.

"Helen, you've got to take time for yourself," the counselors told her. "You have to let go, not see him for a little while."

So, for the first three weeks of July, that's what she did. It was hard for her to break the habit, after crossing the street to her father's house every day for the past few years, for the past decades really, running errands, cleaning, making sure he was taking care of himself, and getting nothing but ire in return. It wasn't easy to stop going over there. Every day she'd look across the street and see a closed-up house. But she resisted the urge.

Finally, she broke. Three weeks had been long enough. Carrying a supply of freshly laundered towels, she crossed the street to her father's house and found him dead in the front hall.

Helen wept over the still-raw memory. Tom and I waited, suffering along with her. She dried her eyes.

The round metal film canister with the original 1910 Edison print inside had somehow gone missing, she said. Tom and she had been unable to locate it in a first cursory search amid the chaos of her father's house.

"I think he loved that movie more than he loved his own grandchildren," Helen said. "I want to find it, because I want to put his ashes in the canister.

It would only be fitting, Helen said, a trace of bitterness playing amid her sadness. "It was the most important thing in his life."

CHAPTER SIX

Every Unhappy Family

Dan Doggett's alma mater

London's lasting shame with many a foul and
midnight murther fed. —Thomas Gray

And if you tease the ragamuffin, gal, you gonna
get kill. —"Murder She Wrote," Chaka Demus & Pliers

Familiarity breeds murder. An axiom of homicide investigation, "Look closest first," means that in any killing, immediate family members, other relatives, co-workers, friends, roommates, acquaintances, and neighbors will receive the brunt of a detective's early attention. Less than a quarter of all murders are committed by strangers. Of all the risk-taking behaviors in which humans indulge, the deadliest may be mundane, mild-mannered propinquity. This fact, along with the news that parents treat ugly children more harshly than pretty ones, establishes the world as a cruel place.

Because of the nature of Aftermath's business, based on indoor cleanup scenes, the company's experience with homicide was skewed even more than usual toward domestic killings. The nuclear family often blows up. Love fails, or, at least, does not prevail. Working at Aftermath fostered a cynical attitude toward the family. I came to regard other people's families as suspect, because I knew at any moment there could be blood on the walls. But it also made me feel a desperate tenderness for my own warm, friendly, nonhomicidal hearth. We were an island surrounded by shipwrecks. Brother kills brother, mother kills infant, husband kills wife and offspring. It made me long for a good old-fashioned Charlie Manson–style home invasion.

Early fall in the London Terrace subdivision of Oak Park, Michigan, failed to show the new development off to its favor. There were no mature trees to burst into color, only thin, shivering striplings, newly planted and, a few of them, already dead. The muddy landscaping was clearly a work in progress.

But to Donal and Marjorie Koss, London Terrace was paradise. They had worked hard for a decade and a half—Donal as a masonry contractor, Marjorie in social services—to be able to afford their dream home. They had moved north, to the "good" side of 8 Mile Road, into a community of upwardly mobile African-Americans just like them. Their quiet, studious daughter, Myesha, had graduated high school as class valedictorian, and had just matriculated into an accelerated premed program at Wayne State, close to home.

When eighteen-year-old Myesha first saw the four-bedroom single-family house of tan brick in London Terrace, she burst into tears. "I never thought we'd live anyplace this nice," she explained to her parents.

The only shadow in the Kosses' blue sky was their son, Myesha's older brother. Twenty-two-year-old Tarell—the family pronounced the name with the accent on the last syllable—had a little problem getting started in life. He couldn't seem to gain traction in anything he tried to do. After getting tossed out of Fenwick by the stern Dominicans who ran the Catholic high school, he meandered through River Forest High with a low C average, never applying himself. He preferred video games, especially racing video games ("Need for Speed: Porsche" was his favorite, ahead of "F1 2000" and "Mobil 1 Rally Championships").

Tarell and Myesha never got along. She always had afterschool jobs and always had money. He never had any. The white pit bull that the family bought as a prize for moving into their dream home was always Myesha's dog. Miss Moneypenny, she called it, after the James Bond secretary, Money for short. When Myesha called her dog, she would say, "Come here, Money," and Tarell made fun of her for sounding white, for acting like a cash-grubbing striver.

What Tarell loved more than anything else was muscle cars. All the Koss men did. Donal had a Victory Red 2002 Corvette parked in the two-car garage of the London Terrace house. Donal's two brothers, Tarell's uncles, had a fleet of five between them, two Corvettes, a cherry GTO, a tricked-out Camaro, and the family pride and joy, a Dodge Viper. For a while there, Tarell had his own, a blue 1998 Camaro with a white hood stripe.

Donal took it away, sold the Camaro right out from under him. He and Marjorie disapproved of the way Tarell spent his time. The last straw was a gun charge Oak Park police leveled against Tarell in the summer of 2005, "unlawful use of a weapon." Police said Tarell had pulled a little .22 automatic on a man he thought owed him fifty dollars on a bet. The charges were dropped and the pistol was destroyed, but from that minute on, Donal kept his son on a short leash.

"You can have a car when you get a job," Donal said. "No job, no car."

Tarell didn't give in. Not right away. It took him six months to gain employment, a half year in which he regularly indulged in screaming fits directed at his parents and his sister. He was nasty, delighting in making fun of his mother's religious devotion, her "Jesus shit," as he called it. It didn't help that during that period, the brilliant Myesha finished her high school education with a bang, delivering a valedictory address at graduation. Tarell didn't show for the ceremony.

He finally did land a job, a miserable gig at a telemarketing firm in Ferndale. He had to take the bus to work, listen to people curse and hang up on him for eight hours, then take the bus back. And still Donal would not relent. He didn't get a car, Donal said, until he proved himself.

It seemed to Tarell as if they were all trying to goad him, trying to get him to explode. Like the jokey card Donal and Marjorie had given him for his twenty-second birthday that September.

"Son," the front of the card read, "I wanted to get you a Jaguar for your birthday . . ." Inside, when you opened it up, was the kicker: ". . . But the darn things need so much exercise and raw meat and they roar really loud in the morning like you wouldn't believe! Happy Birthday anyway!"

Tarell had actually talked to his father before about getting one of the big-engined twelve-cylinder British Jags, an old beater maybe, working on it together, making it purr.

Donal rebuffed him. "You buy a Jaguar, you might as well get a couple, because one of 'em is going to be broke down all the time."

Then, as if twisting the knife, they gave Tarell this mother-fucking birthday card. Making a joke out of his one passion! He literally tossed the stupid card back into Donal's face.

"I'm sick of you all," he screamed, before storming out. But there was nowhere to go. London Terrace was in the middle of nowhere. He didn't even have bus fare. Happy birthday, Tarell.

Tarell's relationship with his family degenerated to the point where Marjorie could not even talk to her own son. She wrote him a careful, two-page letter. "Pray to find your right path," she wrote. "Put your faith in God." Exactly the kind of thing that drove Tarell up the wall.

Two weeks into his telemarketer job, he had had enough. Of

everything. With his first paycheck, he bought a .40-caliber Glock automatic at Chuck's Gun Shop in Riverdale, the single largest supplier of guns traced to crimes in the nation. The law demanded a three-day cooling-off period.

Tarell waited the three days.

After he took the gun home and did what he felt he had to do, he faked a note from Donal on the front door of the London Terrace dream house. "Sorry all, we missed you. My wife has been in a bad car accident. We will be in contact soon."

Then he took Donal's money roll, backed his father's muscle car out of the garage, and drove the red Corvette to Detroit Metro. He parked the car in the airport's short-term parking garage. Packing the Glock in his checked baggage, he watched his green duffel slip through the new X-ray security procedures unscathed. Then he flew to New York's Kennedy Airport, booked another flight to Tampa, and wound up in Florida on a rainy fall morning.

The tempo of Tarell Koss's last hours picked up. In short order he bought a car, paying eleven hundred dollars of his daddy's cash, not for the muscle machine of his dreams, but for a used, piece-of-shit 1992 Toyota Tercel. The most boring, common vehicle on the road. Driving north three hundred miles to the outskirts of Tallahassee, he checked into a motor inn off Interstate 10, again paying cash for the first-floor room. A housekeeper checked his room at eleven o'clock the next morning, opening the door a crack until the security chain stopped her, and saw Tarell lying on the bed. Later that afternoon, the motel manager broke down the door, discovering Tarell Koss dead from a gunshot wound to his head. The Glock lay on the floor beside him.

No suicide note, but fevered scribbles on a pad of paper next to the bed. "You arnt a good Christin evn tho you say you are. You wont hep yor on son." The writing dug angrily into the pad, breaking through the paper in places.

Jacksonville authorities, discovering Tarell's expired Michigan driver's license among his effects, contacted the Oak Park PD, who dutifully drove out to the London Terrace subdivision to notify the family that their eldest had most likely committed suicide in a Florida motel. Officers found Tarell's note on the front door. No one answered their knock, and when they broke in, they were confronted with a bloodbath.

Donal and Marjorie, shot to death in the upstairs master bedroom. Myesha, also shot to death, with Miss Moneypenny dead beside her in her bedroom. Police noted the special ferocity of the attack on Myesha. The killer had pumped thirteen bullets into her body, then used a carpet knife implement to nearly sever her head from her body.

.....

Greg Banach's last name rhymes with *manic*, and he was legendary among Aftermath techs as "Manic Banach," an intense, junk-food-fueled motormouth who reigned at the top of the company's pecking order. He had been with the company for five years, and it showed. Banach could attack any job, deconstructing a badly contaminated room down to its floor beams. He had developed numerous arcane tricks over years of practice. But he appeared brittle around the edges. Despite his bluster ("I don't give a shit—I can stick my face in a bowl of Chinese restaurant rice after scooping up maggots by the handful"), he suffered from bad dreams.

Banach's compact stature contrasted nicely with his partner,

the bearlike Greg Sundberg. To invoke the medieval theory of hu-
mors, if Banach had an excess of bile, then Sundberg was phleg-
matic in the extreme, gentle, happy, slow to antagonize or excite.

I first worked with them on a rifle suicide in Joliet, where
a sixty-five-year-old man shot himself in the basement of his
house. The bullet exited the top of his head, passed upward
through the first floor, and put a hole in the roof. We confronted
a gory, blood-spattered scene in a basement room filled with
huge black plastic fish-breeding tanks. The deceased had bred
exotic tropicals all his life, but was seeing himself shoved out of
the market by the big chain stores.

"This was murder by Wal-Mart," his son-in-law said. "All he
wanted to do in life was raise exotic fish, and when he found
he couldn't do it . . ."

Banach eased the son-in-law back upstairs, away from the
scene. He was a master in handling grief-addled relatives, giving
them firm but unsentimental direction. I could never figure out
where the Aftermath techs, most of them twentysomething ex-
jocks, developed their skills of sympathy and discretion. But it
was true. I was always afraid of saying the wrong thing, but
every one of them appeared to be unerring.

Banach had a mouth like Samson had hair. I once tape-
recorded and transcribed a five-minute stretch of Banach talking.
Within those five minutes he spoke 374 words. I counted.

Working with "Greg and Greg," as the two-man crew of Banach and Sundberg were known, was a whole different experience from Dave and Ryan. Banach had no interest whatsoever in
training me or having me help. I was a sounding board for him. I
sat on the floor of the fish-tank basement and listened to him
bitch about his job.

The money (not enough). The hours (too many). The way

that whenever you plan to do anything with your family, inevitably you get called away on a job. Aftermath techs were on
call twenty-four hours a day.

Every so often he stopped for breath. "Right, Greg?" he'd say,
and Sundberg would say, "Right." Eventually Banach's constant
chorus of "Right, Greg?" got to me. I realized he was actually
talking to himself, asking himself if he was right. The answer always came back in a comforting echo. The first dozen times I
heard the taciturn Sundberg speak, it was always the same word:
"Right."

All the while Banach chattered, he was mopping up a bloodstain the size of a bathtub, from where the rifle suicide had bled
out. There was a special Aftermath technique to cleaning up
blood. The techs tossed a bundle of clean, factory-produced,
white terry-cloth rags into a bucket containing a solution of TR-
32 deodorizer and Thermo-55 disinfectant. They would spray
the bloodstain with Liquid Alive, an enzyme that loosened it,
then fish a rag out of the bucket and go to work. *The rag never
went back into the bucket.* To prevent cross-contamination, the rags
were discarded as soon as they had soaked up all the mess they
could. For the final pass and for general biowashing, an industrial
cleaning product called Spray & Wipe was the tech's best friend.

"Now that's how to clean a fucking bloodstain," as Al
Swearengen said on *Deadwood*. It was a simple process, and Banach managed to keep up a steady stream of verbiage all the
while. After five years on the job, I would have bet he could do it
in his sleep.

Using an aquarium net, dipping into the big black tanks,
Sundberg fished out gray globs of brain matter, a bit of face with
an eyelash still attached and white-gleaming pieces of skull. The
angelfish had eaten the rest.

A week later, when I met Greg and Greg in front of the Koss death house in the London Terrace subdivision, Banach had gotten used to me to the degree he was now throwing an occasional "Right, Gil?" in among the constant dun of "Right, Greg?"

A cold fall evening in suburban Detroit. We were joined by Dan Doggett, a trainee who was preparing to open a California Aftermath office. Balding, intelligent, and earnest, Dan was the husband of longtime Aftermath office manager Nancy Doggett. The other techs all called him Mr. Smithers, off the *Simpsons* character, since he had previously worked nuclear power plants.

"I need all the experience I can get," Doggett said that evening in London Terrace. He was explaining to Banach why he had shown up at the job unannounced.

"I don't mind you coming along," Banach said, his tone indicating that indeed he did mind. "I just wish they would tell me before they hand me a trainee. I think it's right for the crew foreman to know something like that. Right, Gil?"

"Correct, Greg," I barked. I thought he might think I was mocking him, but he was unfazed.

At the end of the Koss driveway neighbors and friends had started a memorial shrine going with a wreath of white roses tied to a driveway pillar. A note with the roses read, "May God bless you all. Your memory will always live on." Down below, a plastic-wrapped bouquet of purple flowers.

As a decorative touch, the builder had topped the driveway pillar with a cherub reading a book (the book of love?) balancing on a mirror ball.

We geared up. The Oak Park CSI team was just leaving. "She's all yours," a detective said to Banach.

"Check," he said. Banach loved hobnobbing with police. His brother and father were both police officers. For a long time he had owned a fully equipped police squad car. The only difference between it and an official black-and-white was its lighting array. As long as Banach didn't utilize blue lights, the squad car was street legal in Illinois. Perversely, Banach kept the backseat filled with stuffed teddy bears.

I always detected a slight pause, a wavering hesitancy, before an Aftermath crew entered a job site. Brief, almost unnoticeable, but it was there. I identified it as the "What fresh hell is this?" moment, after Dorothy Parker's famous line. Just outside the front door of the Koss house, Banach poked his foot at a couple pairs of turned-inside-out green rubber gloves, discarded by police as they left.

We entered to a two-story vaulted foyer. A tan-carpeted stairway curved upward to our right, to the second floor, where Tarell Koss had killed his family. At the bottom of the stairs, the CSI team had casually left a clear plastic garbage bag of bloody latex gloves, soiled paper towels, and ripped-apart body-bag packaging.

"That's nice," Banach said. "Usually they don't pick up after themselves at all. Right, Greg?"

The word brummagem describes cheap and showy ornamentation, the kind that puts mirror balls underneath cherubs. Inside, the Koss house continued the brummagem theme. A chandelier hung from the ceiling of the vaulted foyer, but it was Lucite, not crystal.

In the kitchen, downstairs, a bullet hole had blasted apart a piece of walnut molding. "He started down here," Banach said, "then chased her upstairs."

"How do you know?" I asked.

"Because a cop told me that the parents were already dead when the sister came home. Look, you can track how he followed her."

More bullet holes in the wall of hallway and foyer.

"A couple dozen fucking shell cases," Banach said. "How many shots do you need to kill three people?"

We climbed the stairs. I imagined the sister running, screaming, away from her brother, trying to get to sanctuary. The white pit bull snarling. To the left, Myesha's bedroom. Dog waste piled in the center of an air mattress, where Miss Moneypenny voided when she was shot. Underneath a chest of drawers, a sodden bloodstain the size of a small child gave off a foul odor.

"Smells like death in here," Banach said. "Death city in this motherfucking place."

For a substance so closely associated with a specific shade, human blood spends surprisingly little of its time colored red. Within the body, of course, it cycles through half of its existence in veins as a deep, oxygen-depleted blue. Both the spurting arterial blood and the more sluggish venous fluid turn a bright crimson upon exposure to air. But a pool of blood never remains static. Immediately plasma and platelets begin to separate out. Coagulation paints the stain with prismatic shades of amber, black, and burgundy. Red is only one of blood's colors.

Just above the dark, odorous pool of blood, a poster of the rap star Nelly and another that said "Flower Power" in Day-Glo sixties-style lettering were Scotch-taped to the wall. The vertical blinds clacked in the wind, which soughed through the window from outside. Michigan hurried winter. Stuck along the edge of the mirror, prom and graduation photos showed Myesha, a pretty, heavyset teenager with her scrawny, all-Adam's-apple date.

The CSIs had left behind something else—a thicket of a half-dozen number-two pencils plugged into the bullet holes in the floor.

"They use those to trace trajectories," Sundberg said.

"Correct, Greg," Banach said.

Sighting along the pencils, I traced the bullet trajectory myself. Tarell Koss hated his sister. *Really* hated her. He could spare only a couple shots each for his parents, killed as they slept, but he pumped a dozen into his brainiac, know-it-all, suck-up-to-success sister. "And your little dog too," as the Wicked Witch of the West would say.

"Christ fucking Jesus what a jag-off to shoot the dog," Banach muttered, leaving to check out the other bedrooms.

California, Dan Doggett said, had a lot more regulations than other states. You couldn't just store biomatter in boxes until the waste disposal service picked it up. It had to be refrigerated to forty degrees.

"When I worked at the nuclear plant—"

"Jesus Christ," Banach cut him off. "Again with the fucking Mr. Smithers shit. This isn't a nuclear plant, okay, Dan?"

Chastised, Dan shut up.

To the left off the hallway, opposite the stairs, Tarell's room. Banach said, "It looks like someone ransacked the place, don't it?"

"The cops did," Sundberg said.

The death room of the parents. Banach crossed the room immediately on entering to close the blinds. "You don't want to be the main attraction here."

A massive oak bedstead dominated the room, its four posts topped by foot-round wooden globes.

"That's a crazy bed," Banach said. "That bed is worth nine thousand dollars easy."

"So we're going to try to save it?" Dan asked. I was always impressed by the scrupulous honesty of the Aftermath techs. I never saw anyone even once pocket a "souvenir." On the contrary, I've seen them carefully save many items—a rusty, foul, dust-encrusted heating vent at a "filth job" comes to mind—that anyone else would have tossed immediately away.

"Whoa, look at this," Banach said, fingering a bullet hole torn out of the wood on a sideboard of the bed. "So we can't leave this piece. A family member sees something like this, they'll go fucking bananas."

A huge flat-screen TV entertainment center. DVDs, mostly action but also complete seasons of *The Andy Griffith Show*. On top of the shelving was a plastic dinner plate with an unfinished meal, congealed meat loaf, gravy, and mashed potatoes. And the offending "Jaguar" birthday card to Tarell, carefully saved beside a photo of him in his high school graduation gown. The last good time.

"Look at this, Greg," Banach called. "The bathroom is so big it has a bathroom in it. A bathroom in a bathroom." Declaring that the place still "smelled like stale dog shit," Banach announced he would perform a foolproof trick to deodorize it, the "mouthwash fix."

He flopped a folded towel into the bathtub, then took a three quarters-full bottle of Scope mouthwash and poured the contents onto the towel.

"Tomorrow this place will smell like a crystal clean garden," Banach said. "Mint is the best, but you can use whatever you got."

The crew got down to work. I did the best I could without Banach's direction. Reaching behind a chest of drawers, I picked off a small piece of skull, scalp still attached, that had stuck to the wall.

"You handled that, make sure you put on a new pair of gloves," Banach said as he knelt down in the middle of the room. Using a box cutter, he quickly excised a huge half-moon of bloodstained carpet.

Another trick: Instead of using bioboxes, Greg and Greg tied blue latex gloves ("blue for bio") around the tops of garbage bags with contaminated material in them. They were thus easily distinguished from "GD" bags (for "general debris"). Each biobox or biobag cost $191 to dispose of by a licensed medical waste service, so the distinction was important.

I contented myself with humping the rapidly growing pile of garbage bags down to the bottom of the stairs. On my first trip, I was surprised by a carpenter, banging a sheet of four-by-eight plywood across the front door.

"Who the hell is that hammering?" Banach shouted from up above. He appeared at the top of the stairs. I told him that Donal's brothers had hired a handyman to barricade the door, which the police had broken down upon entering.

"We're going to be fucking trapped in here," Banach laughed.

Sundberg and I dragged two GD bags each through the garage to the driveway, where Greg and Greg's Aftermath truck was parked, its back end open. We tossed the bags in.

"That your truck?" I asked Greg S. about a brand-new Ford Explorer parked alongside the driveway. The SUV had a custom light rack on its roof.

"On the weekends," Sundberg said sheepishly, "we like to fool around with putting light racks, bubble lights, stuff like that, on our vehicles."

I nodded. Sundberg went back upstairs. I humped a half-dozen more GD bags out to the truck and then stood there, taking in the autumn evening. To the southwest, the lights of downtown Detroit glowed pink. I'd heard that every urban area had its own nocturnal light signature, New York white, Los Angeles green. At the end of the driveway, the mirror ball beneath the cherub bibliophile picked up the mercury shine of the streetlamps. The wind picked up, and I thought of the clacking blinds in Myesha's bedroom, above the darkened black stain.

Then that loopy, booming, bottom-heavy bass drum kicked in.

Tell me, tell me, tell me
Oh, who wrote the Book of Love?

CHAPTER SEVEN

The Destroying Angel

The hepatitis C virus

Unless we do something about [Hepatitis C] soon, it will kill more people than AIDS.

—Former U.S. Surgeon General C. Everett Koop

Catch my disease. —Ben Lee

In my first few weeks at Aftermath, I had jumped in with two feet, doing a three-week decomp, a suicide, and a family mass murder in short order. But I was getting ahead of myself. I hadn't gone through the company's training program, except for the informal one that Ryan and Dave gave me on the Milwaukee job. I hadn't studied OSHA directive CPL 02-02, "Enforcement Procedures for the Occupational Exposure to Bloodborne Pathogens." I didn't know what I was doing. Tim told me to come in to the office one morning, and he and Chris would talk about training and next steps.

The Aftermath offices in Plainfield consisted of two twenty-by-twelve office spaces, adjoining two double-story truck bays in the back of the building. The space above the offices was accessible from wooden stairways constructed in the bays, and given over to equipment and chemical storage. Chris and Tim stored some of their toys in the back, too, like an exercise machine and, once winter came, Chris's sleek Wave Rider on its trailer.

I kept my kit bag on metal shelves to one side of the bay. From Ryan and Dave I learned to bring along a gym bag to every job, packed with spare clothes. I chose lightweight exercise garments that would be comfortable under the steamy hazmat suits.

My respirator belonged in the bag, too, although it never seemed to be there when I wanted it, and I packed a pair of running shoes that I used only on jobs. The whole idea was to leave as much of the job behind in the bag as possible, so as not to foul my Extended Stay nest, or cross-contaminate my life in general.

The techs stored their bioboxes in red bins from Medical Waste Solutions, a Gary, Indiana, company that picked up and treated Aftermath's biohazardous waste. Two or three of the bins were parked in a back corner of the truck bay at any one time, with up to seven bioboxes fitted into each of them. Even though Medical Waste Solutions sent a truck around three times a week, the back of the offices sometimes took on the smell of a week-old decomp.

"Keep that door closed!" Raquel Garcia said as the techs wandered in and out from the truck bays, and odor wafted into the front offices.

Raquel supervised the temps who were laboriously compiling and updating a database of every police chief, coroner, medical examiner, and mortician in the country. Waiting for Chris and Tim to free up that fall morning, I listened to the drone of their calls.

"Hello, is this Middleton Police Department? I'm calling from Aftermath, Incorporated, and I'd just like to check on the contact information for the chief of police there. Would that still be Chief Neil White? And is his number the same?"

Over and over, two temp secretaries hard at work, making dozens of calls every hour. Chris and Tim marketed Aftermath aggressively, pursuing contacts and leads at professional conventions of funeral home directors, for example, or annual gatherings of homicide detectives. Civil service rules still prevented many police officers and medical examiners from referring Aftermath directly, but getting the company's name out was vital to its

continued expansion. Many coroners now kept lists of bioremediation companies and would hand out a copy to interested parties,
thereby carefully skirting rules against recommending any individual concern. But it was crucial for Aftermath to be on the list.

"Since a lot of the lists are alphabetical, it helps that we're at
the top," Tim said.

Nancy Doggett, Aftermath's spark plug of an office supervisor, ruled the executive office space. I overheard her speaking on
the phone. "He got caught between the loading dock and a truck
full of lumber," she said. "His head popped like a grapefruit."

Just small talk on an ordinary day at Aftermath, Inc.

The first thing Tim and Chris advised, when I met with them,
was that I should get my shots.

"Make sure your tetanus is up to date," Tim said, "and there's a
hepatitis B vaccine you should get too." The vaccine protects
against a virus formerly known as serum hepatitis. OSHA—the
Occupational Safety and Health Administration, the federal government's largely defanged bureaucracy devoted to workplace
safety—strongly encouraged the shots as a prophylactic measure
for anyone exposed to blood-borne pathogens. This primarily
meant health-care workers, but included bioremediation techs too.

The next day I drove over to a busy clinic in Aurora where Aftermath sent all its personnel for their shots. The nurse practi-
tioner, a sunny young needle-wielder named Jennifer Arroyo,
asked me if I would be working in the health field.

"Have you ever heard of Aftermath?" I asked. "They send
their people here all the time."

She said she hadn't heard of the company. I explained what the
job involved. "They work a lot around blood and body fluids, the
same as you do. Only they sometimes do building demolition at
the same time."

"Wow," Arroyo said. "Well, then it's a really good idea for you to get this vaccine."

She smiled. "But first, I am going to have to give you a little test."

"A blood test?" I asked.

"A pencil-and-paper test," she said.

She wasn't joking. As part of OSHA's certification of Blood-borne Pathogen Training, I was required to examine a fourteen-page booklet entitled "Protect Yourself." The information seemed pitched around a high school level of comprehension, which was okay with me. I hadn't taken a test since my daughter gave me a *Cosmopolitan* quiz on "sexability." The heart of the matter showed up on page five, in large white type on a bright red background. "TREAT ALL BLOOD AND BODY FLUIDS AS POTENTIALLY INFECTIOUS." I thought about the Aftermath jobs I had already been on. Had I done that? I wasn't sure.

OSHA called its infection control measures "Universal Precautions," and the policy was well-recognized by the health-care community. In the absence of a blood test, lacking a certain knowledge of the status of the body fluids you might encounter, you play it safe. Universal Precaution paints the world as a seething, malevolent place, rife with contagion and disease. A wrong step and suddenly you're toe-tagged.

Nurse Arroyo let me study the booklet for fifteen minutes, then gave me a three-page questionnaire. "I've got terrible test anxiety," I told her.

"I'm sure you'll do fine," she said. "It's multiple choice."

And I did do fine, perfect in fact, even though the reward for doing so was to get a hypodermic needle jabbed into my arm. I would need a pair of booster shots, Nurse Arroyo told me, one at six weeks and one at six months. But from the time of the first shot, I would be protected from serum hepatitis. The vaccine was made up of protein cell-coatings from the virus itself, which would stimulate my immune system to produce antigens of its own against the disease.

But it turned out that hep B was not the most lethal worry on the job. Always hovering in the background were the "Twin Reapers," HIV and HCV, the viruses that caused AIDS and hepatitis C, both deadly, both incurable. It was usually impossible to know whether the blood encountered by the techs harbored HIV or HCV, especially since long provirus latency periods sometimes meant that not even the symptomless carriers themselves were aware of the viral presence in their systems.

Several known cases document infection with human immunodeficiency virus, or HIV, the pathogen that causes AIDS, via accidental puncture wounds. Usually these have been nurses or doctors pricked by discarded needles, or "sharps," that were contaminated with the virus. Aftermath techs have participated in the kind of fear and uncertainty faced by health-care professionals, EMT "first-responders," police and fire rescue personnel, and others exposed to HIV in their line of work.

Reggie Fluty, a female police officer on the Laramie, Wyoming, force, was first on the scene of the Matthew Shepard gay-bashing "crucifixion." She found Shepard still alive but covered in blood. The latex gloves issued by the Laramie County Sheriff's Department were faulty and the supply had run out. Fluty said her training told her, "Don't hesitate," so she cleared an airway in Shepard's bloody mouth with her bare hands—hands that were covered in cuts from when she had recently built a shed for her livestock.

A day after Fluty helped rescue Shepard, she was informed that he was HIV positive. Fluty immediately embarked upon a drug regimen and later proved not infected by the virus. Fluty, a mother of two who was later portrayed in the play *The Laramie Project,* said of the incident, "I think it brought home to my girls what their mom does for a living."

"No Aftermath employee has ever been infected on the job," Tim Reifsteck said, after which the normally nonsuperstitious Reifsteck looked around for wood on which to knock.

But there have been close calls. An Aftermath tech I'll refer to as Robby Green worked a week-old decomp in the southwest Chicago suburb of Orland Park. The scene represented what the techs have come to call a "filth job." The term covers the extreme range of decomps, such as the one I encountered in Cudahy, to include locations where deranged individuals have lived and died in unspeakably vile conditions.

The Orland Park scene pushed the limits even in the *Twilight Zone* world of filth jobs. An obese trust-funder named Hennig Geller rarely ventured out of his three-bedroom split-level ranch, preferring instead to remain sequestered within his darkened house, injecting himself with drug cocktails. He gained more and more weight. As his fifty-year-old body rebelled against his lifestyle, his activities gradually became extremely circumscribed, until he was limited to a bedlike couch in his living room, and a narrow, uncluttered path from the living room to the bathroom. Then Geller's heart gave out.

When Robby Green arrived for the cleanup, the three-hundred-pound body had already been removed. But the smell remained overwhelming. Rotted food in the kitchen attracted vermin. The couch in the living room where Geller had ended his days retained the brown, greasy imprint of his body. His heavyweight tread had worn away the carpet on his path to the bathroom, exposing the floor down to splintered wood. In the bathroom itself, two cracked-linoleum shoe prints indicated where Geller had stood to evacuate. The toilet bowl was piled high with feces, which were also spattered around every other inch of floor apart from the two silhouetted footprints.

Heavy-duty demolition work usually comprises a major component of filth jobs, and Orland Park was no different. Body fluids and feces contaminated the carpets, furniture, floorboards, and subflooring. Green's task, to physically remove all contaminated material, meant he would be doing strenuous physical labor around nails, carpet tack strips, and splintered wood, among other sharp objects.

He and his partner got through the job without incident. They transferred the discarded contents of the house into a Dumpster Aftermath had contracted to be parked on the front lawn. They filled biobox after biobox with contaminated material. Green recalled being extra-careful, since this was a rare Aftermath job where they had prior notice about the HIV-positive status of the deceased. Entering the split ranch of Hennig Geller was like walking through a pathogenic minefield.

As Green carried a heavy, rolled-up section of contaminated carpet down the four-step foyer stairs, he tripped and fell. He had wrapped the carpet around several splintered floorboards. As Green stumbled, an evil-looking, gleet-encrusted flooring nail poked out from within the carpet roll and drove its point an inch and a half deep into the flesh of his right calf. Accompanying the physical pain of the puncture wound was the emotional shock of exposure.

"I am fucked, fucked, fucked!" Green screamed to his partner. Blood streaming down his leg, he ran out to the street. A third tech had taken the Aftermath truck for a soda run, so Green ran mindlessly down the block.

"I didn't know where I was going," Green said later. "I was

hoping the truck would be there, or maybe a hospital would suddenly materialize right in the middle of Orland Park."

His partner eventually got Green to a clinic, and as it proved out, he suffered no infection. After blood tests, he received a course of tetanus shots, and that was the extent of his treatment. But the incident affected Green and the other Aftermath techs.

"The thing is, I was being extra-careful, because I knew the guy was a degenerate drug user, and I still got stabbed," Green said. "This is a dangerous job."

....

As intimidating as infection with HIV can be, another danger lurks on Aftermath jobs that could be more lethal still. The viral glycoprotein that causes AIDS cannot survive outside the human body. Exposure to oxygen destroys the virus. Researchers have estimated the risk of being infected with HIV from a single prick with a needle that has been used on an HIV infected person at 1 in 150. In other words, even if the nail that drove itself into Robby Green's thigh had been contaminated with HIV, it still would not have meant an automatic infection. The threat to techs from HIV, while real, is not the main worry going into a job.

HCV is. The hepatitis C virus cannot be destroyed by exposure to air, and is among the most durable and rugged of blood-borne pathogens. Researchers have demonstrated instances where HCV has lain dormant on a surface for a full year, only to become activated and infectious upon exposure to water. HBV can also be communicated through blood-borne contact, but there exists a vaccine against it. HCV cannot be vaccinated against, and there is no cure.

In the "Scuba Doobie-Do" episode on the second season of

CSI, Gil Grissom and Sara Sidle entered a freak show of an apartment, its walls and furnishings completely spattered with blood. The scene of a mass murder? Something more ordinary: nosebleeds. It turned out the resident of the apartment had a grudge against the landlord. A hepatitis C sufferer with weakened nasal capillaries, he could spray blood from his nostrils at will, and did so all over his apartment as a way to make cleanup difficult for the owners of his building.

Assault by a virus-infected nasopharynx might not be the most unlikely scenario on a show that regularly features incredible flights of forensic fantasy. But it ranks memorably high. More startling was the fact that the investigators entered the apartment wearing no protection at all, even after hearing the resident's frank admission of his HCV-positive status. HCV-infected blood can be a deadly hazard. Grissom and Sidle treated it as though it were a weird form of novelty art.

In Stephen King's apocalyptic novel *The Stand*, the microscopic agent that almost ended the world was a protean influenza virus that changed form so often there could be no protection against it. HCV represents the real-life version of King's nightmare. It demonstrates a fiendish changeability. As quickly as the body develops antigens against it, HCV mutates even more rapidly, always staying one step ahead of the immune system's defenses.

Like HIV, the hepatitis C virus would exhibit a terrible form of beauty, were it not for the inconvenient fact that it kills people. It is a marvel of genetic deviousness, so squirrelly it has never been cultured. Researchers did not even pin down HCV's existence until 1989, when the vaccine giant Chiron, Inc., teased out portions of its genome. Before that, the best scientists could do was rather clumsily define the condition it caused as "non-A, non-B hepatitis."

HCV numbers among the lethal mammal-infecting viral

gangbangers of the Flaviviridae family, other members of which spread West Nile, dengue, yellow fever, and hog cholera. Dormancy lends an added dimension to its malevolence. Oftentimes carriers have no idea their blood is contaminated.

Aftermath techs play roulette with hep C. They regularly encounter scenes just as contagious as the ones dealt with so cavalierly by Grissom and Sidle. The progress of hepatitis C is unforgiving. After the HCV virus takes up residence in the host's liver, the organ gamely works overtime trying to expel it. The liver might be "the human body's most amazing machine," as anatomist Henry Gray called it, but HCV is easily the match for it. Under HCV assault the tissues of the liver become inflamed. Abnormal nodules form, and the organ turns cirrhotic and fibrous, the condition familiar from the alcoholic's apothegm, "Hard living leads to a hard liver."

Hep C and alcohol represent only two of the routes to cirrhosis. Industrial toxins such as yellow phosphorus sometimes trigger it, as do dry-cleaning agents, or overdoses of Tylenol. An instant, ready-made recipe for cirrhosis consists of thirty grams of amatoxin-containing mushrooms, including the death cap and the destroying angel. No one survives the destroying angel. Anyone who partakes is dead within twenty-four hours.

Cirrhosis can eventually make the liver act not in its usual role of a filter, but as a plug. The blockage interferes with the vital free flow of the vascular system, and blood begins to back up like water behind a dam. This in turn increases pressure on the arterial walls, which can rupture suddenly, leading to horrific, spurting hemorrhages known to Aftermath techs, who are often called to clean them up, as "bleed-outs."

Dave Creager described a bleed-out scene to me: "You could see exactly what happened through how the blood got painted on the walls. The guy was alone in his house at the time, and you could see it started in the upstairs hallway. All of sudden, like midstep, he was choking on blood. After that starts, you have about thirty seconds before you lose consciousness. He walked down the hall, gushing blood, then took the front stairway down, leaving a trail dripping down the stairs behind him. Then, in the downstairs hall, his gusher really came in, and it hit the hallway wall like a spray in the shape of a fan. He staggered into the bathroom off the downstairs hall. All four walls of the bathroom looked like someone had taken a blood hose and turned it on them. And that's where he collapsed and died."

There are several medical conditions, apart from hepatitis C, that can lead to massive bleed-outs. What did the techs have to protect them against such horror scenes? More to the selfish point, what did I have to protect me? (Me—hey, what about me? Me, me, me!) What fragments could I shore against my ruin?

The hazmat containment suit itself always served as the first line of defense. Every Aftermath truck carried the suits in three color-coded protection levels: white, blue, and yellow. Each provided a more complete line of defense than the shade before it, but the trade-off was discomfort.

"With a blue suit on, you can stand in a bathtub full of water and not get wet," Greg Banach told me. I took a suit back to Extended Stay and tried it, and it worked well enough. The problem was that despite the material's vaunted "breathability," the blue and yellow suits became fiendishly hot, locking in perspiration as they locked out contaminants.

This happened even with the everyday white hazmat suits we wore. Several times I had seen Ryan O'Shea stumble out of a job site after a particularly strenuous bout of demolition, rip off his suit, and pour out a half cup of sweat from it.

The uglier the job, the more affection I developed for the wonder of Tyvek—actually, more properly, Tyvek™. (The papery miracle material is a proprietary product of DuPont, and the company cherishes every ™ and ® it owns.) Like a lot of great discoveries, Tyvek resulted from a mistake. In 1955 a DuPont researcher named Jim White noticed polyethylene fuzz accumulating on the vent of an experimental lab. DuPont patented Tyvek a year later, technically terming it "strong yarn linear polyethylene." The secret was high-density fibers, randomly distributed, which made the material difficult to tear but easy to cut.

The company also quickly patented a "flash-spinning" technique to produce the stuff, and it was off to the consumer races. Starting with book covers, tags, and labels, DuPont manufactured massive sheets of Tyvek at its Richmond factory, softening it for drapability, applying "corona treatment" finishes so that it could be printed upon. Builders wrapped virtually every new home in America in Christo-like swaths of Tyvek, and FedEx adopted the material for its ubiquitous overnight delivery envelopes.

"Did you know that during the early 1980s," I asked Greg Banach as we geared up one day, "the governments of Haiti and Costa Rica printed their currency on Tyvek?"

"No," Banach said, sounding peeved. "I didn't know that, and I didn't want to know it, and fuck you for knowing it."

"You ought to stop doing that," Greg Sundberg said.

"What?"

"That fun fact shit. It only annoys him."

Banach had probably climbed in and out of a few thousand or more Tyvek hazmat suits than I had, so the thrill was long gone for him. But the material wrapped me on every job I did for Aftermath. The process of suiting up itself became pleasurably ritualized, as though I were putting on a uniform. At one job site in a southern Chicago suburb, a small flock of spectator children called out, "Hey, spaceman! Hey, spaceman!" whenever I waddled down the sidewalk in my Tyvek.

Me and Dick Cheney. The vice president's support team brings along his level-four chemical-biological Tyvek hazmat containment suit wherever he goes. The suit is never more than a few yards away from him, which ought to make you feel either more secure or more nervous, depending on your feelings toward Mr. Cheney. What might he know that we don't?

Since HBV, HIV, and HCV pathogens do not readily atomize to become airborne, the breathing apparatus we were fitted with wasn't for protection so much as odor elimination. The 3M respirator I eventually settled on—an eighty-nine-dollar 6800 model, size medium, with a full clear-acrylic face-piece—wouldn't shield me from poisonous gases, but it did prevent the sledgehammer odor of decomposed human flesh from slamming into my sinus membranes quite so heavily. It also muffled speech, which meant it had the added benefit of shutting up chatty, job-shadowing crime writers. Sometimes I think the techs stuck one on my face just to prevent me from asking stupid questions.

Most of the techs cheerfully accepted my presence, laughing at my occasional bouts of nausea but solicitously attending to my safety and education. Ryan and Dave, especially, took me on as a pupil. They called me Hollywood in tribute to my screenwriting aspirations.

"Hollywood loves his Tyvek," Dave said, watching me pull on a second hazmat suit over my first one, just to make sure.

Apart from Tyvek and 3M, other brand names had worked their way into Aftermath culture. Purell hand sanitizer. All the techs drank Red Bull energy drink, although Dave liked Mountain Dew MDX, too, when he could get it.

But the holiest of holies was Axe body spray. Not just any body spray. Had to be Axe. Though not, strictly speaking, a disease-prevention measure, body spray did inoculate against certain cases of social ridicule and ostracism. Axe was as integral to an Aftermath job site as Tyvek or 3M. The easy-grip molding and the matte-black paint job on the container made it fit right in, as though it belonged in any manly man's tool chest.

"If I could haul a barrel of this stuff in the back of the truck, I would," Ryan said, liberally dousing himself in Axe Apollo after stripping off his Tyvek. Apollo was the preferred scent among the techs ("Modern and sexy, for the clean-cut, superjock effect. Put down the medicine ball and spray on some of this stuff.") but I've seen Kilo, Voodoo, and Phoenix in play too. I've since discovered that Axe enjoys a raging popularity among high-school-aged males everywhere, which helps explain the obsession on the part of the techs. They were, most of them, just out of high school themselves.

The scent began to have a Pavlovian effect on me. I'd smell it, and my mouth would start to water, since application of body spray was always the last step before heading to Thursday's, Olive Garden, Hooters, or whatever the day's chain restaurant of choice might be.

"I'm hungry as a hostage," Dave would say, moving within a cloud of Axe. "Let's go eat."

....

There were down times. What the French call *longueurs*. Jobs would come in barrages and then abruptly cease altogether. The Naperville Extended Stay became my purgatory.

Hanging around the room got a lot worse after I went to a job at another Extended Stay franchise, this one north of Chicago, in a suburb called Hoffman Estates. A five-day decomp, a lodger who had expired in the bathtub. As soon as we entered the room, I realized I was in trouble.

It was my same room, with the same flecked-green rug, same plum-colored easy chair on gold casters, same locked-down TV remote, same energy-saving light fixtures. I had a surreal sense of disorientation, like maybe *I* had died in the bathtub of the sad room with the green carpet, and what I was doing now was all some kind of *Jacob's Ladder*-style postmortem brain spasm.

Everything was exactly the same as my room in Naperville, with the kitchenette to the left and the bathroom to the right. Only, this bathroom had a collection of dead man's change on the sink countertop and an ugly crucifix shaped stain on the bathtub. I could see where the head had rested, and where his arm had flopped over the side of the modular tub, because those areas were traced in black, like the shadow on the shroud of Turin.

When I came back from that job, I couldn't work up a lot of enthusiasm for Room 112, my home away from home. I avoided its horror at the Steak 'n Shake next door (which, frankly, harbored its own brand of fluorescent, oleaginous horror). Finally, I had no choice. I had to face the green-flecked carpet.

"I've been meaning to ask, Mr. Reavill, what is the work you do?" Mahesh, the always-smiling thumbs-up hotel manager, said to me as I trudged through the lobby.

I have always hated telling people I am a writer. It strikes me as pretentious. I told him I was in town on business.

"I see the men come by in the truck all the time," Mahesh said. "Aftermath? What is that?"

Ryan and Dave had been stopped by the Illinois state patrol for running commercial plates on an unmarked truck. So Chris and Tim ordered all the Aftermath vehicles into a local sign-painting shop for a logo to be applied once again to their front drivers' doors. Smaller this time than the twelve-inch Day-Glo of Aftermath's early days. Discreet white block letters on the dark blue paint of the trucks, with the 1-800-TRAGEDY number printed below.

"We clean up crime scenes," I said.

Mahesh appeared nonplussed. "You clean up crime?"

"After people die, or after they are killed, the coroner takes away the body, but sometimes there's blood and body fluids left. We clean that up."

He lost his smile. An expression crossed his face that I have come to call the Universal Yuck. Usually accompanied by head-shaking and sometimes, not always, succeeded by a kind of un-willing, mesmerized-by-the-cobra fascination.

"That must be very upsetting work," Mahesh said. "I feel for you, my friend."

His ritual thumbs-up appeared forced after that, and I started to avoid him by slipping into the place via the back door.

During a slow period I took a commuter train into Chicago, to the Museum of Science and Industry, to see the Body Worlds exhibit then enjoying equal parts notoriety and acclaim. A bus-man's holiday—a phrase that has unfortunately become out of date, meaning a vacation where one indulges in activities similar to one's work. A specific part of the exhibit interested me.

Body Worlds was a traveling show of preserved human bod-ies dissected and displayed for observation, some of them posed in *Totentanz* postures—a flayed cadaver carrying the sheath of its own skin, for example, one dribbling a basketball, one rearing on a dissected horse. With seventeen million visitors worldwide, Body Worlds has been enormously successful, spawning several imitators. The show has also attracted its share of controversy over the source of the cadavers on display, and the seemliness of displaying them at all.

Aftermath dealt with the human body as a leaky, messy vessel. What I saw at the museum was the quite the opposite: the body drained, preserved, tidied up. Both Aftermath and Body Worlds, however, dealt with the body as object. Both were rationalist enterprises, respectful of the dead, but declining to participate in the more extreme forms of sentimentality associated with dying.

Anatomist Gunther von Hagens began plasticizing bodies for exhibition in the mid-1990s, in his post at the Institute for Anatomy at the University of Heidelberg. The first show he mounted—the word was especially fitting—in Japan in 1997, drew 2.5 million visitors, and encouraged him to bring Body Worlds home to Germany. He was unprepared for the ugly reception. Media, church leaders, and ethicists attacked Body Worlds as exploitative and disrespectful. The shadow of the Nazi lampshade fell over the exhibit.

In London, also, there was controversy, but in America von Hagens's creations were received with the kind of wonder and appreciation I saw all around me at the museum that day. Von Hagens "plastinated" his cadavers, meaning he injected polymers into the cells. The resulting plastinates have long been used as teaching tools in medical schools, but von Hagens was a pioneer in popularizing (he calls it "democratizing") gross anatomy.

As I passed among the displays, I felt the shock and awe of the other visitors. I particularly liked the bloodred spider work of the vascular studies, which resembled nets in which to catch the human form, or circuitry for the body electric.

I recalled that Herman Webster Mudgett, aka H. H. Holmes, America's first widely publicized serial killer, was an anatomist who murdered many of his victims in his self-designed "Murder

Castle," just a few blocks west of the Museum of Science and Industry grounds. Active during the 1893 World's Columbian Exhibition, held lakeside right where the museum was located, Holmes no doubt would have delighted in Body Worlds. As would the mad taxidermist Ed Gein.

There is a logical disconnect to that observation—Ed Gein might have liked Oreo cookies, too, and that doesn't make Oreos any less sweet—but the Body Worlds enterprise struck, for me, a very familiar note. As with Aftermath, Incorporated, Body Worlds was fraught—fraught with controversy, fraught with emotion, fraught as a target for moral judgments.

To a certain degree, von Hagens played off these controversies. As a publicity stunt for one of his German shows, he seated a plastinated pregnant woman on a public bus and sent her around Berlin. He was quoted as saying he wanted to display plastinates having sex, and that he wanted to pose one crucified on a cross.

I found what I was looking for in an obscure corner of the exhibition featuring stand-alone displays of various human organs, diseased and otherwise. The smoker's lungs looked predictably gruesome, with an ugly gray-black color to them that couldn't be good. But I stopped longest in front of a case containing two cirrhotic livers.

Even if you had never seen a human liver close up, and thus had nothing to compare, you would instantly conclude there was something irregular about these gnarled, fibrous hunks of diseased tissue. The specimens were under glass, but I wouldn't have wanted to touch them. They looked like tree boles. A small card informed me that one of the organs had come out of the body of a hep C sufferer, and I idly wondered how von Hagens and his technicians decontaminated the thing. Some thinly sliced cross-sections of cirrhotic livers, looking equally gnarly, were mounted nearby.

So this was what all the fuss was about. Without HCV, without HIV, Aftermath could be just another janitorial concern—no hazmat suits, no $191-a-throw bioboxes, no Universal Precautions. Staring into the vitrine at a squamous football twice the size of a healthy liver, all warty and misshapen, the fuss seemed worth it.

I took the train back to the western suburbs, passing directly by, I noticed, the site of the infamous Murder Castle of Dr. H. H. Holmes, on Sixty-third Street just east of Halsted.

....

The *math* in the word *aftermath* comes from the Old English word *maeth*, "to mow grass." So *aftermath* means "after the mowing." During New York City's wilding craze during the 1980s, the teenage gangs of muggers used to call their victims "wheat," boasting that they mowed them down just like threshing machines. "All flesh is grass," said the cold-eyed prophet Isaiah. "Surely the people are grass."

A public relations consultant gave Aftermath its name in 1999. "After Crime Clean-Up" wasn't working out. The company's original name ruffled too many feathers and raised too many hackles. So Chris and Tim looked for something that was suggestive, but not too suggestive.

Sterling, Illinois, where they both grew up, and where they had known each other since second grade, used to bill itself as "the Hardware Capital of the World," due to the dominating presence of smokestack industries such as Northwestern Steel & Wire, Lawrence Brothers Hardware, and the Wahl Clipper Corporation.

The working-class town situated itself at a falls on the Rock River in Whiteside County, a hundred miles west of Chicago. Tim's father worked at Northwestern Steel and Wire, the medium-sized steel stamping mill that employed half the town's fifteen thousand population, turning out "long products": nails, wire, bar, rod, and structural rolling.

Northwestern Steel had a contentious labor history throughout the period when Chris and Tim were growing up. Steel dumping from Japan and Germany killed off a huge sector of the American steel industry, and Northwestern saw a three-year strike in the mid-1980s. Reorganized, the mill started up again, and after high school Tim followed his dad by taking a job there. He signed up for the hottest, dirtiest, most dangerous work in the place, the furnace crew, because it paid a premium.

His experiment with old-style rust-belt industry ended badly. Working one day under a smelting furnace in temperatures of two hundred or three hundred degrees Fahrenheit, issued wooden shoes because the soles of ordinary boots melted in those conditions, Tim dehydrated and collapsed. They had to carry him out.

"I quit and never went back," he said. He began searching for ways to make a living that were not quite so nineteenth century. When he partnered with Chris after college in their newspaper subscription business, they were both just treading water. They looked for the next big thing, and thought they might have found it in Aftermath.

Working side by side on cleanup jobs, they discovered that they complemented each other. "Chris and I are so different," Tim said. "He's a lot more outgoing, and I'm almost a recluse. So he does the talking, the letter writing. I became more the technical person."

In the beginning and all through the late 1990s, the company was still very much a break-even proposition. "Anything we made we plowed straight back into it," Chris said.

Tim still lived in the apartment building across the street from that first volunteer cleanup job. As Aftermath ramped up, he focused on his work, played basketball a couple times a week, and maintained a nonexistent social profile. He kept his apartment decor bachelor Spartan: a big-screen TV, a recliner, a bed. That was it. Most nights, he wouldn't get home until ten or eleven.

A dark-haired beauty in her early twenties lived across the street. She would watch Tim walk out to the mailbox in his basketball sweats, his head down. "I could see straight through to his house when he opened the door and I could see he didn't have any furniture," the former Sara Muse recalled.

"She used to look in my window," Tim said, laughing.

Sara had to make all the moves for the shy, work-focused Tim. She got his phone number from her mother, who worked as a concierge at his apartment building. She called him up and the two began dating. In the winter of 2001, they were married.

Sara Reifsteck bears a striking resemblance to the movie actress Selma Blair. She was there at the company's beginning, through its days as a shaky start-up concern, with the subscription service covering the bills. She watched Aftermath expand and develop into a commercial juggernaut, and is quite obviously proud of her husband's success. When she worked for a time in the Aftermath office, her firsthand experience of man's inhumanity to man affected her deeply.

"I couldn't believe all the horrible measures people dream up to do to one another," she says. "It's like a whole spectrum of hurt."

Sara and Tim have two young sons and want more children, but divulge information about Aftermath to their kids on a strict need-to-know basis. "I own a restoration company," Tim tells them.

Chris Wilson was also raising a child, his daughter, Avarie, who was a little older than Tim's sons. Chris treaded lightly around her as far as Aftermath was concerned. "My daddy's an entrepreneur" was how Avarie used to describe her father's work. In a kindergarten exercise about what parents do, she told her teacher that her father was a "house fixer." Now in second grade, she's graduated to a more graphic description: "My dad helps people out in places where there's blood."

The summer that I started job-shadowing at Aftermath, Chris married a gorgeous, vivacious Plano woman named Kelli McKirgan. She also worked in the Aftermath office for a stretch, and brought friends and relatives into the company. A natural-born saleswoman, she traveled for weeks at a time cold-calling sheriff's departments, funeral homes, and medical examiners' offices, publicizing Aftermath's services. I asked her if the nature of the job affected her.

"It's not the murder and suicides or the brains on the wall that got to me," Kelli said. "It's the filth jobs. We're not talking about people who don't dust their mantels. When I was in the office a job came in that was a mother who was deranged, who never cleaned up at all. She kept all her garbage like it was precious."

Kelli saw the site photos that the techs brought back to the office. Used toilet paper, diapers, and feminine products trailed out of the house's single bathroom. The woman's toddlers fingerpainted with their own feces.

"Okay, I know you're abnormal," Kelli said. "But why can't you flush the toilet?"

Both Aftermath wives worry less now that their husbands have taken on executive duties and don't work in the field much anymore. They cited what they see as Chris and Tim's commitment to tech safety and the principle of Universal Precaution.

"If I had a bioremediation job, and I was looking to hire a company," Kelli said, "and I looked outside and I saw someone roll up in a pickup truck with a bunch of mops and a Wet Vac, you know, like a mom-and-pop deal, I swear I would lock the door and hide. Knowing what I know, you can't do this job safely with half measures."

The Moral Compass thought that all of Aftermath's precautions were at least partly a way to pump up their fees.

"You know what a cynic is?" I asked her after she trotted this theory by me on the phone.

"I know what a cynic is," she said.

"A cynic takes the worst aspect of anything and says that it's the most important aspect," I said.

"Okay, okay," she said.

"You haven't been on these jobs," I told her.

"No, I haven't. But you shouldn't swallow the company line whole hog either. Maybe you need some cynicism. You're supposed to be a journalist."

Insurance companies were indeed on the warpath against Aftermath. Cleaning up a trauma scene Chris and Tim's way cost a certain amount of money. But as far as I could see, the company's precise and careful adherence to Universal Precaution was the only approach that made any sense.

At a family mass murder that occurred less than two miles

from my Naperville Extended Stay refuge soon after I took up residence, I got a firsthand look at another company's approach to bioremediation.

.....

Daniel Dellenbach's friends sometimes invoked the word *charismatic* to describe the twenty-eight-year-old health insurance broker. At parties and gatherings around the chain restaurants and brewpubs of the suburbs west of Chicago, he proved himself smart, witty, and charming, a blond, blue-eyed ex-jock with a steady girlfriend and a good job.

Most of his Chicago friends didn't know about Dellenbach's dark side. During an aborted stint at the University of Iowa in Iowa City, a nineteen-year-old Dellenbach had been arrested on four counts of what the Iowa authorities characterized as "home invasions"—he burglarized residences when the occupants were at home. He washed out of juvenile boot camp on that charge, and wound up spending six months in county. He moved back to his parents' house in the western Chicago suburb of Wheaton, but was nabbed on a shoplifting charge.

Slowly, in his mid-twenties, Danny seemed to right himself. His parents, Ginny and Mark, were rock-solid citizens, both hard workers, with Ginny relishing her reputation as the Betty Crocker of Wheaton. Life went especially well for his sister, Brenda, who was waiting tables at Famous Dave's barbecue when she met a young, moneyed computer programmer named Rick Livolsi. He was on a date with another woman and Brenda served them at the restaurant. Surreptitiously asking for the pretty brunette waitress's phone number, Livolsi took Brenda out the next night. They'd been together ever since, even eloping to Reno on a lark.

Danny looked on, a shade jealous, as his formerly prospect-less younger sister moved into a four-bedroom luxury home in a Warrenville golf course development called River Hills. The "river" was the West Branch of the DuPage, just a creek really, but it flowed prettily through the eighteen-hole golf course just outside Brenda and Rick's front door. Livolsi wrote code, owned his own computer programming firm, and drove a Ferrari.

When anyone asked Rick how things were going, or even casually inquired how he was doing, he always gave the same answer. "Life is superb."

Danny had a good job, too, but he couldn't quite keep up with his sister and her new husband. He sold health insurance to the small and medium-sized companies that flocked to Chicago's exurban ring cities. With commissions and bonuses, he pulled down fifty thousand dollars. But it never seemed enough. Part of the reason was the nasty little blow habit he'd imported from Iowa City. He needed money, always more money. Money to party, and money to be the life of the party, which was what Danny Dellenbach always had to be.

So he began to steal from his family. Nickel-and-dime stuff at first. His father, a straight-arrow salesman who loved to help his wife throw large family gatherings, noticed that a few hundred dollars would be missing from his bureau-top money cache every time Danny attended one. Then Rick missed a twenty-five-hundred-dollar money roll left in an upstairs bedroom, also just after Danny visited.

Finally, his sister, Brenda, confronted him. "I know what you're doing, Danny," she said. "I know about the credit cards."

Dellenbach had secretly filled out applications for plastic using his parents' sterling credit information. A lot of plastic. All told, he had taken eighty thousand dollars from credit card companies, mostly under Mark's and Ginny's names, but using Brenda's and Rick's too. Most of the money went up his nose and those of his friends. The life of the party.

"Don't tell Dad," Danny pleaded to Brenda. He knew what straight-arrow Mark would do if he found out his son was engaged in credit card fraud. There wouldn't be any coddling or restitution. Mark would go straight to the authorities, and Danny would go straight back inside.

It wasn't a prospect Danny felt he could face. County time was well known to be the worst kind of time there is, and Danny's six-month stretch in Johnson County Correctional had been pure despair—horrific food, crazed fellow inmates, constant sexual and physical intimidation. He could not do another stretch. He just couldn't.

"Danny, of course I have to tell Daddy," Brenda said. "How can I not? Where would you come up with the money to make this good?"

Danny erupted. "You tell Dad and I will kill you," he shouted. "I'll kill you, Brenda!"

His sister wasn't much impressed. She'd had put up with a lifetime of Danny's tantrums. Danny could scream all he wanted. Brenda kept finding new fraudulent credit lines, and the losses spiraled upward. The other shoe was going to have to drop.

Danny made his move before Brenda made hers. The night before Halloween, two weeks after he had made his threat against his sister's life, Danny Dellenbach again showed up at River Hills. The gatehouse guard recognized Danny and waved him through. He pulled up in front of Rick and Brenda's palatial home on the golf course. Rick was at his computer in the den, and Brenda was watching television. Danny angrily confronted his sister. Had she told their father about his misdeeds?

Before Brenda could answer, Danny was on her, pummeling her with his fists. She yelled for Rick, and Danny seized a fireplace poker and smashed her across the head with it. When Rick rushed into the living room, Danny swung the poker at him, catching the left side of his head, knocking him to the carpet. Danny hit Rick a dozen more times, great pummeling blows, any one of which could have been fatal. Then, with his brother-in-law safely out of the way, he returned to his sister and finished her off, slugging her savagely to cave in her skull.

He wasn't through yet. As it neared midnight on Halloween eve, he drove back out through the River Hills gate and headed east to his parents' modest ranch home in Wheaton. He carried a .44-caliber pistol. Mark and Ginny were asleep. Danny fired at his father first, a kill shot to the left side of the back of the head. His mother woke at the explosion and turned groggily to face her son. Danny shot her once in the face, the bullet entering above her lip and ripping through her cerebral cortex.

Intent on his business now, Danny loaded up the bodies into his leased Toyota sedan. It could not have been anything else but a terrifyingly surreal drive, heading through the suburban landscape on a raw October night, his dead mother and father in his trunk. Through the River Hills gate with his dark cargo, back to Brenda and Rick in the bloody living room. He carried his parents' bodies into the house.

A Halloween family reunion of sorts. They were all together now. Danny didn't belabor the festivities. He drove past the gatehouse guard for the fourth and final time that night, and caught the last flight out of O'Hare to Atlanta.

....

It was an odd coincidence, and it made me feel all the more like the Angel of Death, for a quadruple murder to happen in my immediate neighborhood just as I was waiting around for quadruple murders to happen in my immediate neighborhood. If I stepped out of my Extended Stay enclave to the hotel parking lot, I could look across the interstate and see the southernmost edge of the River Hills development. It was that near.

I called the Aftermath office and spoke to Chris. "You guys didn't have anything to do with this, did you?"

Chris laughed. "Too close for comfort, huh?"

"I just feel strange," I said. "The worst news possible for somebody else is pretty good news for me."

"You'll get used to it," Chris said. "And you'll be glad to hear that it looks as though we are going to get both jobs, the sister's place near you, and the parents' house over in Wheaton."

Danny Dellenbach fled to a friend's house in Atlanta, but then turned around the next day and flew back into Chicago. He rented a car and drove toward his old college neighborhood in Iowa City. An Iowa state patrol officer spotted him on I-80 and pulled him over. He was in jail in Iowa on a felony menacing charge (his threat against his sister), fighting extradition to Illinois.

Aftermath did not get the contract to clean up the horror house at River Hills. At the last minute, the insurance company stepped in and insisted the family hire a cheaper national cleaning chain to do the work. Greg and Greg did clean the master bedroom where Mark and Ginny Dellenbach died in Wheaton, and I briefly visited the scene.

It was, as Aftermath jobs went, very simple. Head wounds tend to bleed out egregiously, but the blood was contained on the mattress. The carpet did not have to be ripped up. The bullet that killed Mark Dellenbach tore a groove in the headboard, so the techs packed that up and disposed of that too.

"Even though it's not biowaste, it's a reminder of what happened," Sundberg said. "So we take it."

Through a real estate contact of Tim's, I got to witness how bioremediation was done on the cheap. I accompanied Tim and his contact on a walking tour of Brenda and Rick's house, which would soon come on the market. (Full-disclosure laws in Illinois and many other states meant that the seller would have to tell prospective buyers about the homicides that had transpired there.)

We first visited the River Hills house when the cleaning firm was in the midst of its job. Thick green hoses trailed in through the front door, leading back to a steam unit in the cleaning company's van.

"Look at that," Tim said, disbelieving. "They're just using a regular carpet cleaner."

Rentable at grocery and hardware stores, ordinary steam cleaning units work by blasting steam through carpet or furniture fabric. A return line extracts the resulting waste and feeds it back to a storage receptacle. The problem was, the approach doesn't work all that well.

"A steam cleaner like that leaves all sorts of stuff behind," Tim said. "I'd say maybe fifty percent of the biomatter will still be in there when they are finished. It'll look clean, but it won't be clean."

And indeed, when I returned to the house with Tim's contact for a walk-through two months after the murders, I could smell it. The living room exuded the musty, telltale funk of decayed blood. I put my face to the carpet and inhaled, and could smell it even more sharply.

"Jesus," Tim's contact said, "it smells bad in here."

"Smell the carpet," I said.

"Um, no thanks," she said.

She wasn't going to be showing the Livolsi house anytime soon. As the bioremediation industry grew, operators were pouring into the field, some of them much less professional than Aftermath. Chris and Tim always told me they wanted to be the Cadillac of bioremediation, the standard against which the field was measured. I initially dismissed such talk as promotional inflation. But after catching a glimpse of the competition at the Livolsi house, and after putting in my time on the HCV killing fields of suburban Chicago, I became a believer.

Man Versus Machine

A turbofan

There are many accounts, uniformly incomplete, of what it is like to die slowly. But there is no information at all about what it is like to die suddenly and violently. —Martin Amis, "The Last Days of Muhammad Atta"

The machine easily masters the grim and the dumb. —Marshall McLuhan

Just before 9:00 A.M. on Sunday, January 16, 2005, Continental's Flight 1515 boarded passengers at El Paso International for a jump across the state to Houston's George Bush Intercontinental Airport. The first officer for the flight did the traditional walk-around inspection of the Boeing 737-500 and noticed a small patch of fluid ("probably oil," he reported) beneath the right-hand number two engine. Boeing designers had slung the big GE CFM-56-3 power plant, with its distinctive "hamster pouch" look, underneath the wing, making the engine easily accessible for maintenance.

While passengers were still shuffling into the fuselage above them for the morning flight, a crew from El Paso's Julie's Aircraft Service motored out to the site, popped the cowling off number two, and tried to trace the source of the leak. The day kicked off clear and sunlit, with scattered clouds beginning to blow in off a gentle north wind. This early it was still cold, forty degrees. Among the four-person maintenance team was a sixty-four-year-old veteran A & P (for "airframe and power plant") mechanic named Donald Gene Buchanan.

Buchanan had worked around the Boeing 737 for the whole of his career. He encountered the plane more than any other, which was not surprising given the 737's wide use. It's the most popular jet in the sky, making up a quarter of the worldwide commercial fleet, and the best-selling craft in the history of commercial aviation. For its distinctive stubby fuselage, pilots nicknamed the plane "Baby Boeing," "Guppy," or "Fluf" (for "fat little ugly fella"). About 1,250 are airborne at any given moment, and a 737 takes off or lands somewhere in the world every five seconds. So for Buchanan, troubleshooting on Flight 1515 that Sunday morning, it was as though he were working with an old friend.

He had earned a reputation around the El Paso airport as a better-than-competent mechanic and an all-around good guy. For Paisanos al Rescate ("countrymen to the rescue"), a group that drops water to immigrants making the dangerous border crossing in the deserts to the west of El Paso, Buchanan completely rebuilt a thirty-year-old Cessna. The mechanics and airline personnel made up his true family—he lived alone in a small house along Interstate 10, a couple miles south of the airport. He liked to feed the neighborhood cats.

The maintenance team communicated with the cockpit flight crew via a plug-in intercom system used whenever passengers were boarded. The crew noted that the leak occurred only when the engine ran.

"We need you to idle number two and then spool it up to seventy percent," Buchanan told the captain.

Normally, screamingly noisy engine "run-ups" such as this took place some distance from the terminal, at a designated location two and a half miles away in the holding bay of Runway 22. In rare instances—and this was one—National Transportation Safety Board rules allowed a pilot to test the engine while still at the gate.

General Electric's CFM-56-3 delivers anywhere from 18,500

to 23,500 pounds of thrust from a seven-foot-long, two-ton engine. In the engine's nose, behind its domelike elliptical spinner, a five-foot compressor-mounted turbofan with thirty-eight aluminum alloy blades whirls around at 3,600 rpm. The fan provides the engine with the enormous gulps of air required for operation. It is a collection of spinning knives.

At 9:23 A.M. Donald Gene Buchanan straightened up from a crouch and took a single step into the "hazard zone," a fore-engine vortex of air intake. Tornado-strength suction lifted him off his feet and fed him into the turbofan, which effectively atomized his flesh. Blood and viscera coated the compressors, and a small amount burped out of the engine's stern exhaust.

The passengers already onboard heard a thump and felt a shudder rock the plane.

"Something's caught in the engine," the copilot said to the captain, who immediately powered down.

The death of Donald Gene Buchanan represented a particularly extreme example of a particularly violent category of Aftermath jobs, which are given the tag around the company of "man versus machine."

Even taking into account the remorseless nature of human violence, a certain ferocity marks man-versus-machine deaths. The CFM-56-3 turbofan that killed Buchanan had no feeling about the event one way or another. The holy trinity of crime—means, motive, opportunity—in this case lacked that vital central element. Man-versus-machine deaths embodied the "motiveless malignity" that Coleridge identified in Iago. There was no motive, and thus no appeal to fairness, balance, or revenge in the event's aftermath. Nature was implacable, which lent an element of frustration and despair to those affected by it.

What killed Buchanan was easy enough to plumb. But it's not a whodunit, it's a whatdunit. Here's what killed him:

$$\dot{m} = \rho \cdot V \cdot A$$

In other words, a law of physics perpetrated the crime, a mass flow rate (density × velocity × flow area) that created the kind of vacuum so abhorrent to nature, which rushed in to fill it, sweeping the body of a man to his death.

"Engine ingestion" is the euphemism aviation regulators apply to such incidents, mostly with geese, ducks, or other birds. Human engine ingestion had happened previously, but always in hangars, never before at a gate. A reporter from the *El Paso Times* interviewed several of the 144 passengers who were on Flight 1515 as it sat on the tarmac. They expressed sympathy for the victim's family. Some of them were visibly shaken, although Continental later said none actually saw the accident occur. "Many of the passengers," the paper noted dryly, "were interested in finding out what happened, where their luggage was, and how they could book another flight."

What Continental Airlines had on its hands in El Paso was an unusable $10 million aircraft engine. Simply discarding the four-thousand-pound behemoth wasn't an option. Continental equipment managers needed it cleaned.

Who could they call?

....

Humans help machines kill other humans all the time. Firearms themselves could be seen as machines ("deflagration-driven compressed-gas projectile devices"), and as the NRA never tires of informing the public, "guns don't kill people, people kill people." The most common category of accidental death in Western countries is man versus machine: automobile fatalities, most of them attributable to human error.

Perhaps there was a human element to blame in the death on the El Paso runway. Aviation safety experts (and some retired mechanics) had proposed lanyards and safety harnesses for everyone working around the low-to-the-ground 737 engines, but none of the maintenance team wore them that day. The engine run-up might have been more safely accomplished away from the terminal, out on runway twenty-two.

But human error didn't change the fundamental truth that lurked behind most of the man-versus-machine deaths I encountered at Aftermath. The universe is a cold, impersonal place. A jet engine, a piston, a drill press, a smelter, a printing press, an industrial saw, a bakery oven, or a high-pressure hose can work with horrific effect on human flesh. There were industrial machines that I learned could kill but of which I had never heard, like cleatformers, screw feeder presenters, and automatic pin chucks. The human body, piece of wondrous work that it is, turns out to be not much of a match for a twenty-five-ton cold-stamping roller.

At times man and machine colluded to spectacular and gruesome effect. In an industrial district on Chicago's South Side, a depressed filing clerk at a Health and Human Services record-storage facility took the same route to work every day. He emerged from the orange line train station and walked west, by a long line of garages and warehouses. He passed the open bay of Tivoli Foundry, a white metal scrap concern that collected copper, zinc, tin, lead, and aluminum from discarded materials, melted them down, and sold the resulting ingots on the world market.

Workers at the foundry grew accustomed to seeing the clerk

pass. They fired the Cowper stoves and blast furnaces early in the morning, so fires were already glowing, booming out a sound much like an aircraft engine, by the time the streets filled with workers heading for their nine-to-fives (in Chicago, it was more often eight-to-fours). Scavengers showed up with commandeered shopping carts full of scrap. The darkened interior of the huge foundry space provided a good contrast to the spark and radiance of liquid copper. Passersby were always stopping to stare at the cherry-orange glow of the molten metal as it was fed down the feeder troughs into the molds.

Most of the smelting buckets were small affairs, capable of only limited volumes, but the aluminum cauldron was huge, resembling a cement mixer without wheels. The aluminum glowed a pretty yellow-silver shade as it melted. Catwalks and railings surrounded the smelter.

On a spring morning the government clerk didn't walk by Tivoli Foundry the way he usually did. This time he walked directly into the bay. The employees were caught unawares. Nobody said as much as a "Can I help you?" as the clerk made a beeline for the aluminum smelter, mounted a ladder, and jumped in. Upon being immersed in the superheated metal, the wet flesh of the clerk's body literally exploded, covering the immediate area around the smelter with smoking, burning gore. Much of the biomatter had simply been vaporized, Hiroshima style.

The resulting cleanup was a fifty-hour Aftermath job.

As grisly and massive as Tivoli was, it could not match the three weeks Ryan O'Shea and Dave Creager spent in a Continental warehouse outside Kansas City, painstakingly cleaning the six thousand parts of Flight 1515's number two engine.

At first, the enormity of the job stymied them. They had never attempted anything so complex. But after the first few days, the task proceeded in an orderly, if time-consuming, fashion. Aviation engineers dismantled the engine, passing each non-electronic part to Ryan and Dave for cleaning. They would first immerse the part in a bath of Microban and Liquid Alive enzyme cleaners (they had brought whole barrels down to Kansas City for the job), loosening the particles of biomatter. The techs then cleaned the part by brush or cloth. The electronic parts were cleaned separately, without liquid immersion, with a half mile of wire scrubbed inch by inch.

"You couldn't really see that much that looked like remains," Dave said. "Some parts of the engine were just, like, coated with a film. But we didn't find bone bits or hair or anything we thought we'd find."

Brain matter was always the most difficult substance to deal with, and it was notorious for its strange ability to spatter across unlikely distances. Bobby Hargis, a motorcycle cop riding in John F. Kennedy's motorcade in Dallas, was struck by the president's brain matter, and told the Warren Commission, "It wasn't really blood—it was a kind of bloody water."

When "bloody water" dries, it becomes difficult to remove. Aftermath techs usually attack it with putty scrapers ("It's a bitch to get up," a tech told me). At Kansas City, Ryan and Dave had to be more delicate. The brain matter had worked impossibly into nooks and crannies of the engine parts. Repeated soakings could not remove the sticky crust. It took several applications of Spray & Wipe, coupled with the enzyme cleaners, to do the trick.

The whole job took more than two weeks, over 150 man hours. Ryan and Dave worked weekends, with the Continental engineers and mechanics putting in overtime ("Work hard—fly right" was the current slogan of the airline, an improvement on the former one: "We really move our tail for you"). But at the end of that time, the company had what it wanted: a $10 million engine, ready to be reassembled and recycled.

Somewhere in the skies, one of the proud 737 "birds with the golden tail" flies with its engine parts cleaned and restored to use courtesy of Aftermath, Inc.

CHAPTER EIGHT

Ageless

Twice a day

There is but one truly serious philosophical problem and that is suicide.—Albert Camus

Why wait?—Suicide note of George Eastman, inventor and founder of Kodak

"You haven't really worked this job until you've done two things," Greg Banach told me, " —a three-week decomp and a shotgun suicide. If you can handle both of those, you can handle anything."

Three-week decomp, check. The horror show in Cudahy. It was midfall when I headed west from Chicago to Irene, a small hamlet on the southwestern edge of Rockford, Illinois. Dave Creager and Ryan O'Shea were going to clean up the shotgun suicide of an eighteen-year-old boy.

I followed along behind Dave and Ryan's truck in my rental. Catching a ride with me was Joe Halverson, a new company hire. The Aftermath recruitment process tended to stick close to home. Halverson was an old high school friend of Chris Wilson's wife, Kelli.

The route MapQuested through an exurban landscape of farmland and "build-aways," modern single-family homes on wooded lots or on the edge of cornfields. We humped over a set of railroad tracks and drove along a row of older homes, ill-favored, asphalt-shingled residences and preranch bungalows, with a few prefabs and double-wides thrown among them. Ryan and Dave turned right up the steep driveway of a small, two-story brown house set above the road beneath a stand of sugar maples. Joe and I parked across the street and walked over.

No one around.

"The brother was supposed to be here," Dave said, and he got on the cell phone.

Across the driveway ten yards away, a neighbor's house was also shut up tight, the only sign of movement in a rabbit hutch in the backyard. Standing next to the Aftermath truck I again felt like the Angel of Death. It was sunset, and the eastern sky had already lost the light. Beyond the line of maples, dark brown evening fields lay fallow. Across the street were the parking lots and grounds of a sprawling new high school, most of the building itself invisible behind the huge brick box of its gymnasium.

A small bowlegged man in his mid-twenties started up the steep asphalt driveway toward us. He staggered visibly, wobbly on his feet. Dave walked down to meet him. The rest of us suited up as Dave spoke to the big brother of the deceased.

"I guess the place is empty," Ryan said. "The family is at the brother's down the way."

He looked over at the blank-windowed house. "This is going to be bad," he said.

Dave and the brother headed to the back door. The brother said that the homeowner's insurance policy was inside. "Would you mind if you went in there with me?" he asked Dave. "I don't want to go back into the house alone."

Dave said he would. I watched the two of them enter. There was something wrong with the brother. He couldn't walk, and his speech was a mumble, almost incoherent. He appeared ill.

Dave and the brother came out, and the brother walked unsteadily down the driveway and headed up the road.

"Upstairs," Dave said. "A twelve-gauge."

"All right," Ryan said.

"What was wrong with the brother?" I asked Dave. "Was he drunk or something?"

"No," Dave said.

"Well, he looked like he was having trouble walking."

Dave looked over at me. "What you were seeing was some-one deep in grief."

Right. Of course. I guess I had never really seen it before, at least not to that degree. When my father died there were tears and a solemn air to the gathering, but he had had a stroke in the years previous, so his death did not come as a lightning bolt to my family, the way the unexpected death of a child would. The brother of the suicide had been a total wreck.

The first clue was a tooth, its bloody root still attached, on the narrow carpeted stairway to the second floor. We had come in through the gloomy, darkened first floor of the house, clut-tered with the ordinary clothes-food-paper disarray of day-to-day life. Off the hall between the living room and kitchen, the stair-way took two sharp turns upward, and we found the tooth on the first landing. There didn't seem to be any possible way it could have been blown down two ninety-degree turns to wind up where it was.

Ryan picked it up with a gloved hand. "How do you think that got there?" I said.

"You won't believe what a shotgun blast does to the human body," Ryan said. "I've seen blood spatter make it around cor-ners. It can get under doorways and bounce up over the wall in the next room even when a door is closed."

At the second landing I could look upward and see the first markings of blood on the white ceiling above me. The second floor was really a narrow finished attic with two rooms in line, and the suicide had pulled the trigger in the first of these. A blood bomb had exploded, and all four walls were covered with gore. Most of it had been propelled toward us, toward the stairway and a closet next to it.

The floor of the closet was piled high with clothes, a teenage boy's disorganization. The blast covered the clothes on the floor and those hanging against the back wall of the closet in a fine spray of blood, small clots of brain, and bits of skull and flesh. Amid the mess I saw something I had never seen exposed before, the white matter of the brain, nerve bundles surrounded by sheaths of fatty myelin, normally tucked away deep inside solid gray masses of neurons.

"Jesus," Ryan said about the contents of the closet. "All that is going to have to be bio."

He stepped back to survey the room, putting on his Gil Grissom hat.

"He must have been sitting here," he said, "with the shotgun on the bed beside him, pointing into the room. So he reaches back for the trigger—look, you can see where the stock kicked back and hit the wall."

A small oval dent in the white wallboard. Was it really the mark of a gun stock? Or maybe the indentation of a heel, made long before in happier times. I couldn't really tell, but no one said anything to challenge Ryan. It was just a theory, a "what happened here?" puzzle game. Nothing rode on the truth of it. The Winnebago County coroner had already ruled the death a suicide.

Everything in the room was absorbent: the wallboard, the stain-mottled indoor-outdoor carpeting, the green coverlet on the bed, even the bulletin board tacked up on the door to the second bedroom. Ryan opened the door.

"I don't see anything in here," he said, after kneeling to make an inspection. "I think this room is okay."

Dave clattered upstairs, somehow managing to carry three bioboxes at once. He put them on the floor in one of the few clean areas in the room. "I'll start on the closet," Ryan said. He reached down for a load of the blood-soaked clothing and transferred it to one of the boxes. "Jesus, there's a lot of shit in here."

"It all has to go?" Dave asked.

"It all has to go," Ryan said.

Joe, the rookie, worked the far side of the bed. "What's this here?" he said, bending down, and then straightened up with a yelp.

"It's an eye," he said.

A folded-over flap of blood-soaked skin had concealed a perfectly preserved eyeball, its blue iris flecked with black, and the eggshell-colored sclera still moist with lubricant tears. All four of us gathered around to examine the gory remnant.

"Holy shit," Joe said. "What's the last thing it saw, do you think?"

Dave and Ryan laughed, teasing him. "You yelled like a little girl when you picked it up," Dave said.

"It just surprised me, that's all," Joe said.

Although they politely put up with my retching, reeling, and absurd fluttering "vapors," the techs were merciless with each other. Any hitch, any startle, any bridling, was grounds for instant derision. The merest shiver in the face of the most horrific scene imaginable was seized upon with a happy viciousness.

To a man, all of the techs swore they had never vomited on the job. They called it "pulling a Bushie," after the memorable incident in 1992 when the first President Bush puked all over the prime minister of Japan at a state dinner. It didn't matter how much brain matter splattered on the wall, how gross the maggot mass was, how clogged the toilet was with feces—none of the techs had ever pulled a Bushie.

Joe placed the eyeball reverently into a biobox. On the bed, at the source of the 270-degree fan of blood spatter, lay an open high school yearbook.

"That's him," Dave said, gesturing to a black-and-white junior-class portrait of a thin, pimply youth with a thick sweep of hair across his forehead—some of the hair, and some of the forehead, that we were picking out of the wallboard of the room we were in.

"Hell, that photo's not that bad," Ryan said. "Not bad enough to kill yourself over."

Dave laughed and moved to help him clean out the blood-painted clothes closet.

....

From studies by psychiatric academics, we know quite a bit about suicide. Generalities about gender, for example: that men tend to use guns and women pills, or that men complete suicide at a rate higher than women, but women attempt suicide more often than men.

As one would expect, Monday is the prime day for suicide. Less explicable is that Tuesday has the lowest rate (all the gloom got sucked up by Monday?). There is a peak in the spring (April is the cruelest month) and a lesser one in September. Late afternoon, from 4:00 to 6:00 P.M., are the witching hours, actually better and, in this context, more ironically known as Happy Hour. The fourth through the sixth of every month are the favored dates. Astrological signs do not correlate to rates of attempt or completion. Capable as always, firstborns complete more often.

Contrary to popular belief, there is no spike in the suicide rate on Christmas, although the rate of attempts rises slightly on New Year's Day. People don't commit suicide any more often on their birthdays.

There is also no rise in suicide rates on Super Bowl Sunday, during the final game of the World Cup or seventh game of the World Series, but the rate dropped in Scotland for four years after the Scottish team was in the Cup. Athletes involved in club sports commit suicide less often when compared to the general population.

We know that of all countries in the world Hungary has the highest suicide rate, and Hungarian immigrants have elevated rates when compared to other immigrant groups. Communist and Socialist voters in European elections have higher rates, capitalists lower. In the UK, suicides drop after a Labor victory and rise after a Conservative victory. Suicide rates in October and November 1929, around the time of the "Black Tuesday" stock market crash, were the same as those in January, February, and September of that year, belying the widely held image of defenestrating stockbrokers.

Physicians (especially female doctors and psychiatric specialists) and pharmacists have elevated rates, while dentists and nurses have average rates. Rates of farmers and foresters are higher than truckers. Rates of police are second highest of all professions, behind only psychiatrists (physician, heal thyself).

We know other odd, mostly gender-related differences. In cases of suicide by burning, men tend to pour flammable liquid on themselves and light it, while women tend to walk into already-burning flames; men more often shoot themselves in the head, women in the torso. Male jumpers leap from greater heights than female ones. Men, uncommunicative brutes that they are, don't leave as many suicide notes as women. Males with tattoos are more likely to use guns as their method of choice, and brown-eyed males are more likely to use hanging or poison.

Some suicide studies slice the demographic pie a bit thin: Self-poisoners in Sri Lanka, for example, do not exhibit gender differences when they choose between rat poison or agrochemicals. Other studies provide eureka moments. From the 1970s on, when emission standards cut the amount of carbon monoxide in car exhaust, that method of suicide drastically declined.

Prisoners in World War II concentration camps had lower suicide rates than the general population, which leads some theorists to conclude that the act may be less likely if you can blame exterior factors for your misery. Suicide rates for teenagers in the West went up as their living conditions improved. A free press increases the homicide rate but not the suicide rate. Finally, psychiatrists and sociologists who study suicides (suicidologists) have an elevated rate of self-authored death.

As a species, we are not alone. Monkeys self-mutilate in zoo environments, sometimes to the point of death. Iguanas can end their own lives. Likewise certain protozoa, pink bollworm moths, and some species of wasps and butterflies.

It was difficult, within the circumstances of their jobs, for Aftermath techs to maintain the type of equilibrium set forth in the ancient Hindu text of the Bhagavad Gita: "For certain is death for the born, and certain is birth for the dead—therefore over the inevitable thou shouldst not grieve."

"Easy for him to say" was the only way to respond after a hard day at Aftermath, Inc. Chris and Tim employed a grief counselor,

Mary Ellen McSherry, to help clients deal with the bruising new reality of their lives. But McSherry often wound up counseling Chris and Tim and other Aftermath employees too. Sometimes it wasn't just a job, it was an emotional obstacle course.

This was especially true for the thorny existential puzzle of self-authored death. Chris and Tim were consistently amazed at the persistence of the destructive impulse in suicides. Time and again, they witnessed victims who had purchased a gun at Wal-Mart, sat out the obligatory three-day waiting period, then taken the weapon home and killed themselves.

"The Wal-Mart sales slip is right there on the bed beside the body," Chris said, shaking his head in disbelief.

Method was sometimes as strange as timing. Not once, but several times, Aftermath cleaned up after suicides that were committed by table saw, including one instance where the victim purchased the saw, lugged it home, set it up, and then lowered his neck onto the blade.

The most destructive suicide Chris and Tim witnessed was that of a young man, a federal judge's son, who was despondent over failing his EMT test. The son went into the family's basement den, placed a stick of dynamite between his teeth, and lit the fuse. The wick had a three- or four-second burn time, so the judge's son had a long moment to meditate on what was about to happen.

The blast destroyed the den, rendering the wood-paneled walls into shredded, paper-thin remnants of what they had once been. A patch of the ceiling blew upward and punched a hole into the first floor, where the judge was sitting at the kitchen table having coffee. The body itself sat intact on the couch in the basement—but only from the waist down. The rest of the remains were exploded around the room, teeth embedded in the shredded paneling.

Chris and Tim learned a good lesson about the media on the

day they came to clean up the basement den. Because of the prominence of the father, the son's suicide was big local news. They had difficulty maneuvering their Aftermath van close to the house, amid the cluster of news vans, each with a satellite antenna reaching toward the sky. It was August 31, 1997. They had to get a police officer to guide the Aftermath van through to the site.

But suddenly, the tight traffic jam of media broke up. Reporters ran to their vehicles and roared away, burning rubber. It was like magic. Only a few minutes had passed since the house had been besieged, ringed by news vans. Now the street was empty. What had happened?

The news that Princess Di had perished in an automobile crash suddenly dwarfed all other events. The horrific suicide of a federal judge's son was no longer deemed newsworthy.

"We never took media attention quite as seriously ever again," Wilson says. "Something else can always come along and bump you off the front page."

....

Ryan and Dave were careful and sedulous workers. The suicide in Irene required ripping up large sections of carpet and the removal of a six-by-eight patch of wallboard. The shotgun pellets had torn up the drywall to the degree that it proved impossible to clean. Ryan and Dave did most of the work, a machine well-oiled by years of experience.

It helped that they had been friends for so long. They had developed an almost psychic way of working with one another, in which many of their exchanges consisted of two or three clipped words, mystifying to outsiders but perfectly intelligible to the two of them.

Joe and I tried gamely to keep up. I didn't feel I had the expertise to work on the more contaminated areas of the bedroom. Mostly I lugged the bioboxes down to the truck.

Dave recalled that his first job, after hiring on at Aftermath, was a shotgun suicide.

"A trial by fire," he said. "The guy shot himself in the bathroom, and there were big chunks of brain in the bathtub. Once you smell brain, you never forget it."

"You can smell it right now," Ryan said from his position on his knees behind the blood-spattered bed. There was an acidic, chemical stench to the air, lingering underneath the more metallic smell of blood.

Ryan and Dave came out of Plano and Sandwich, two small farmland communities next door to each other west of Chicago. Dave grew up in a semirural setting, in a house built by his grandfather and set on nine acres of fields and woodland. He was all of nine years old when his father, Jack, who worked at the big Caterpillar factory in Montgomery, stuck him in the driver's seat of a bulldozer and unleashed him to help carve out a fishing lake in front of the house.

He had two brothers, Jack Junior and Paul, and a host of cousins. It was an idyllic childhood, especially during the thick Illinois summers in the gently rolling farm country outside of Sandwich. "We were outside playing around all day and most of the night," Dave said. "The only way you could get us inside the house was by yelling the word *food*." Willie Nelson has a song, "Me and Paul," about his longtime drummer Paul English, but Dave always liked to think it referred to him and his brother.

"I was always a motorhead," Dave said. When he was eighteen years old he got his hands on a '67 Corvette with a 454 "big block" engine that he rebuilt, racing with it on the deserted back roads between Sandwich and Plano. He was too restless for the classroom ("You can't put me behind a door—I just go crazy") but excelled in school sports, especially basketball and football, with a 4.4 forty and a forty-inch vertical jump. The local school teams were called the Plano Reapers.

At a party in high school he encountered a rock-solid middle linebacker named Ryan O'Shea—nicknamed "Rhino" as a play on "Ryan O," but also because of his pugnacious personality ("O'Shit" was a less kind sobriquet). Ryan had not quite as stable a family environment as Dave did.

"I watched my dad get last rites five times," Ryan said, whose remote, disengaged father suffered under assaults from diabetes, cancer, and a dangerous blood condition called a cholesterol shower. Perhaps because of his terrible health problems (exacerbated by a three-pack-a-day cigarette habit), Ray O'Shea was never there for his son. He passed on his Irish bulldoggedness to Ryan, but not much else.

"He's always going to be my father but he's not really my dad," Ryan said.

Instead, Ryan helped circle the wagons around a tight family trio of his mother, Pat, and his little sister, Katy. He started working at thirteen, picking apples in an orchard, and never stopped. He tried to convince his mother to accept reality and divorce his father. The family endured foreclosure and food stamps, and Ryan had the humiliation of his schoolmates holding fund-raisers to benefit the O'Sheas.

He took out all his aggressions on the football field. The middle linebacker position is among the most brutal in sports, but Ryan excelled at it, copping All-State Illinois honors two years in a row. He and Dave buddied up at parties, off-road excursions, and motorcycle forays around Plano and Sandwich.

"It wasn't a good weekend unless we got into at least one fist-fight," Dave said.

"Crazy Creager and Psycho Rhino," Ryan recalled. "We scared the shit out of people."

After aborted careers in collegiate sports, Dave and Ryan both wound up driving trucks for a living. They both made plans to go into law enforcement. Dave took his civil service exam and was waiting to hear from several small police departments in the western suburbs. But when his cousin Kelli told him about a company called Aftermath that her boyfriend, Chris Wilson, had started, Dave signed on. He eventually brought his best buddy Ryan in, and the two partnered up on the company's second crew.

Soon afterward, Dave's brother Paul came near to death in a downstate auto accident. "I went down there to visit him, because he was too banged up for them to move," Dave said. "He was in real bad shape."

Peel back the surface of any life and you will find plenty of misery, misery enough to provide abundant reason to do yourself in. "For the world's more full of weeping," says Yeats, "than you can understand."

Somehow, humans locate powers of endurance that are beyond all expectation, beyond all logic. Despite onslaughts of tragedy in their lives that would bring lesser creatures to their knees, when I suggested that for some people suicide was a way out, both Ryan and Dave dismissed the idea out of hand.

"Maybe life just gets too much to bear for people at times," I said.

"Fuck that," Ryan said. "You fight." He wasn't laying judgments on anyone who killed himself, he said, but it would never be something he'd even consider. He mentioned his love for his sister, Katy, who he had watched out for since she was a small child. He rolled up his sleeve to show me a matching tattoo, a Celtic heart that he got after Katy got hers.

"I've seen too much of what you leave behind when you kill yourself," Dave said. "Families are just distraught. You can't know the pain you cause."

Dave Creager took me on a driving tour of Sandwich and Plano, pointing out the stomping grounds of his high school years with Ryan. The countryside around that area, on the far eastern rim of the Great Plains, has a fecund, Land-of-Goshen feel to it. It reminded me a lot of the dairy country of central Wisconsin, where I grew up. Dave drove by both their houses, and the football fields of their glory days. A prominent feature of the tour was the package goods store where the two of them had bought their beer during high school.

We wound up at Judy and Jack Creager's spacious home, set amid the woods in a development near Silver Springs State Park. Jack Creager was just headed out to pick up Dave's brother Paul, now fully recovered from his accident after years of painful rehab. I caught some of the capable, hardworking nature of the son in the figure of the father. I sat at the kitchen table with Dave's mother, Judy, talking about the family and about Paul's near-death experience.

"The thing I like about Aftermath," Judy said, "is that it's helping people out in their time of need." It was a theme sounded by virtually everyone who worked at Aftermath, from Chris and Tim to the secretaries and the techs. They all cited altruism as the aspect of the job that was most satisfying to them. There was a Midwestern earnestness to the claim that made it difficult to second-guess.

Judy worked out of her home, distributing a vitamin product called Ageless. She had experienced four successive bouts of cancer, she said, and she had become very proactive about her health. She offered me some free samples of Ageless, a blueberry-colored liquid supplement that needed to be refrigerated, and that you were supposed to take twice a day. Judy looked healthy to me, cheerful in the face of what was, for her, lately, a hard life.

The assaults of accident and the depredations of disease, the seemingly endless catalog of pain and tragedy and mortality that I encountered at Aftermath, had begun to affect my outlook. I saw death everywhere. It was like buying a new car and then suddenly seeing that make and model constantly, whereas before you hadn't ever noticed it. Like a lot of people, I managed to block out as much misery as I could.

I thought about that as I drove away from meeting Dave Creager's family, a large plastic baggie full of vitamin supplements next to me on the seat of the rental.

Ageless.

In Greek mythology the gods granted a prophetess named Sybil a single wish. She asked for eternal life. It didn't turn out well. She forgot to ask for eternal youth. As she lived on and on over years and decades and centuries, her body shriveled with age until she was the size of a grasshopper. The high priests put Sybil in a bottle and kept her at the oracle at Delphi.

Once in a while someone would open the bottle and ask, "What do you want, Sybil?" And Sybil would say, "I want to die."

Agelessness is the true deep-down wish of every human being, and the one thing denied us. I didn't know about the health-giving properties of the juice Dave's mom was distributing, but they got the name right, at least. I popped a couple open and sucked them down.

....

A month after the cleanup of the shotgun suicide in Irene, I re-turned to talk to the family. I hesitated before calling them, since the death of the boy in the yearbook was still no doubt fresh in their minds and hearts, but when I spoke with the victim's older brother—Pat Sullivan, the one who had difficulty walking when I first saw him—he said he wanted to talk about the tragedy. He told me he'd prefer I not come by the house, so as not to upset his grieving parents. We spoke at a restaurant in a strip mall outside Rockford, and over coffee, Pat Sullivan told me about his brother.

Sean Sullivan lived most of his life in the cramped asphalt-shingled house catercorner from John F. Kennedy High School, which opened when Sean was eleven. He did well in grade school and junior high in nearby Rockford, and was marked as an intelligent, sensitive kid.

His father, a long-distance trucker, was absent for long stretches, and his mother worked swing shift as a nurse's aide in Rockford. The Sullivan boys, Pat, Wesley, and Sean, were alone at home for the majority of evenings in the week, but their aunt and uncle, their mother's brother and his wife, lived five hundred yards away, in a farmhouse back off the main road.

"Sean started acting funny the last year of junior high, when he was in ninth grade," Pat told me.

"Funny how?"

"It was hard to put your finger on," Pat said. "I thought he was smoking too much pot. I smoked myself, so I wasn't one to talk. But the older he got, the stranger he got."

When he hit high school, troops of kids would be in the house every day after school. Pat had already graduated and wasn't around much. "But I know they had a lot of bong parties, stuff like that."

Heavy pot use has been associated with depression in teenagers, and Sean began to withdraw socially as a junior. His grades fell off. Then, on the afternoon of his seventeenth birthday, he tried to rob a branch bank in Rockford with an orange-colored water pistol.

"That was a real wake-up call. He just shoved a note at the teller, like 'I'm here to rob you, give me the money.' But he was just a scrawny kid with a toy gun. The cops overreacted. They tackled him, put him on the ground, and handcuffed him and all that."

Sean spent six months in psychiatric lock-down at Singer Health Center.

"I think it made him worse," Pat said. "He went around to all his old high school friends, and told them that they were the ones who made him crazy."

Sean missed school a lot in his senior year, and it got him in trouble with his social workers, since he was on supervised release from the hospital. He stayed home and watched television—or watched the high school across the street from an upstairs bedroom window.

"I saw him like that a lot," Pat said. "He'd be curled up on Wesley's bed, because Wes had a better view of the school. I'd ask him, 'Sean, what are you staring at?' He'd say, 'Nothing.'"

The social work team at family services talked about returning Sean to the hospital. He spent a lot of time paging through the yearbook from his junior year, defacing some of the pictures with a ballpoint pen.

What do you want, Sean? I want to die.

Sean had his choice of firearms in the house. Every fall, all the members of the Sullivan family, men and women both, hunted deer in the corn-rich Illinois farmland around Irene.

"Sean usually used a shotgun with a solid slug," Pat said, "A lot of times some of the places we hunted were near houses, and we didn't want a rifle to take somebody out."

The day Sean shot himself he came home from school, a rare day when he attended all his classes. He was alone in the house. He opened the yearbook on his own bed, not Wesley's, but he could still see a small slice of the high school from where he sat. He positioned himself awkwardly, twisting around so the blast was directed into the room and not toward the wall behind him. Loading the 12-gauge with a single shell of heavyweight buckshot, he pressed the trigger with the thumb of his right hand.

Pat Sullivan didn't speak about the details of his brother's death. Talking about it made him feel lost, he said. He asked me what he could have done that was different.

"We tried to get him help," he said. As we said good-bye, he attempted some rueful gallows humor.

"You know, we all would have preferred it if Sean had used the real gun to hold up the bank, and used the toy gun if he wanted to hurt himself. That would have been a lot better, don't you think?"

CHAPTER NINE

The Long Pig

The Chicago Rippers in the news

In most cities, there's a constant battle between good and evil. In Chicago, everybody gets along. —Lenny Bruce

You get more with a smile and a gun than you get with just a smile. —Al Capone

The mantle of criminality has never rested easily on Chicago's big shoulders, but the city has never quite been able to shake it off either. "The only completely corrupt city in America," one disgusted nineteenth-century reformer labeled the town. A hundred years later, when the son of a prominent political family wedded the daughter of a reputed mob figure, the *Chicago Tribune* did not deign to cover the marriage in its society pages. The story belonged exclusively to the paper's crime reporter.

Going to Chicago for crime is like going to Vichy for the waters, and it's been that way since the very beginning. The city erected itself on shifting foundations, both of the moral and physical variety. Building on the wet, sandy soil of a marsh on the southern rim of Lake Michigan, the city founders encountered immediate difficulties with multistory structures. They couldn't build big. The foundations sank into the quagmire. An apt metaphor, some said, for the moral quicksand that was the city. Feet of clay, indeed. Chicago existed on extremely rickety stilts.

For a time, engineers attempted a bizarre, makeshift solution. They created an understory and jacked the buildings downtown ten feet off the ground. From their sidewalks, respectable denizens of Chicago could gaze downward into that shadowy understory—a muddy nineteenth-century twilight zone of beams and trusses, exposed to the elements and representing an open hunting ground for legions of cutpurses, thieves, and thugs. The city fostered a criminal underworld that was among the most vibrant, enduring, and powerful in history.

Architects eventually solved the sinking-foundation problem, to the degree that Chicago became the skyscraper capital of the world. Shifting moral foundations were harder to address. The criminal element that graduated from downtown's strange architectural understory went on to rule the city and, eventually, the whole country, dictating legislation, electing presidents, and gorging themselves on graft.

During the Roaring Twenties and the gift to the underworld that was Prohibition, there were 976 gang murders in Chicago,* a death count not really matched until the crack epidemic hit Los Angeles sixty years later. Chicago in the twenties had its own "death corner"—at Oak and Milton, where thirty-eight victims were gunned down in a single year—and even its own "dead man's tree"—a poplar on Loomis Street upon which Black Hand assassins would pin the names of their victims.

From the feuding fiefdoms of Al Capone, Johnny Torrio, and Dion O'Banion grew Chicago's reputation as a gangster's paradise. Out of the Prohibition-era mob grew the "Outfit," a vast organized-crime enterprise that spread its influence to Las Vegas, Los Angeles, and beyond. The Outfit helped put Harry S Truman and John F. Kennedy into the White House, and infiltrated countless sectors of American business, industry and, society. "Crime does not pay," stated *Mad* magazine's Alfred E. Neuman,

---

* Only two of these murders yielded criminal convictions.

"as well as politics." Chicago has shown that when you can combine both, you've got it made.

Growing up, I felt Chicago hovering at the southern border of my world, gazing evilly at me like the eye of Sauron. Both my parents were raised there, and for a while in my preteen years I spent my summers at my grandmother's house just off Kedzie Avenue, while my mother took her education degree at the University of Chicago. Later, in the early 1980s, I returned to the city with a girlfriend who attended medical school there.

My first Aftermath jobs had all been in the western suburbs, or farther afield, in Wisconsin, Iowa, and Michigan. But a few weeks before Thanksgiving, Chris Wilson called and asked if I wanted to join a crew of techs on a body removal in Chicago.

"Sure," I said. "But I thought Aftermath didn't do that much business in Cook County."

In most Aftermath jobs, the actual corpus has long since disappeared, bagged up and hauled off by personnel from the medical examiner's office. Many times days or even weeks might go by before the Aftermath crew arrives to deal with the human leftovers.

But in a small percentage of cases, the local authorities contract Aftermath to remove the body. Several jurisdictions, primarily smaller ones in the western suburban counties of Kane, Kendall, and Will, occasionally use the company's removal services. Cook County, which embraces the city of Chicago proper, normally does not.

"The job's east of Archer Avenue, a Polish neighborhood, Pope John Paul Drive," Chris said. "Joe and Kyle are going to handle it."

Joe Halverson had completed his training under Ryan and Dave. He and his stepbrother, Kyle Brown, now made up the

third and least senior crew operating out of the corporate headquarters of Aftermath. Usually they inherited the jobs no other
techs wanted—the jail cell cleanups and squad-car details that Aftermath does gratis, as goodwill gestures to the sheriff's offices
and police departments it depends on for referrals.

Joe was stocky and heavy-browed, with short-cropped brown
hair, while Kyle was slight, more wiry. But they both displayed a
sunny bonhomie in the face of the gloomy and macabre situa-
tions they encountered on the job. I rode in their GM, an older
truck they were assigned as the low men on the Aftermath totem
pole. Joe installed a green canvas camp chair between the front
seats and I squeezed in.

I asked them if they'd ever done a body removal. "No, and
I'm not wild about it," Joe said. "I don't like to see dead people.
I'm fine with cleaning up their fluids and their shit and their
odor, but even at funerals, I don't like to see dead bodies."

"But we'll take it," Kyle said. "We haven't been working that
much lately."

At Aftermath, techs are treated not as salaried employees but
as subcontractors. If a tech doesn't work, he doesn't get paid.
Jobs go first to the top crew, Greg and Greg; next to the second
crew, Ryan and Dave; and last to Joe and Kyle. Aftermath charges
its clients (or their insurance carriers) $250 an hour for the ser-
vices of their trained biohazardous waste removal teams. The
techs receive varying portions of that, from $80 an hour for Ba-
nach down to $45 each for Joe and Kyle. There are extra charges
for materials, demolition, chemicals, and biobox disposal. There
have been instances in which a Wet Vac used on a job had to be
discarded because it had been contaminated with biomatter.
That and similar charges go on the client's invoice.

"Where are we going?" I asked.

"Over by Comiskey," Joe said.

I corrected him. "You mean 'U.S. Cellular Field.'" The trend to corporate branding had hit the venerable Comiskey Park, home of the Chicago White Sox.

"I hate that stupid name," Joe said. The White Sox were deep into their World Series run, and the tech crews split evenly along north (the Cubs) and south (the Sox) lines.

I examined the MapQuest printout that Nancy Doggett, the Aftermath office manager, had prepared for Kyle and Joe. Computer-generated mapping has been a godsend for far-flung service companies such as Aftermath, but the crews widely deride the maps and directions as difficult to follow and oftentimes downright wrong. On a MapQuest printout "2.8 miles" sometimes turned out to be more like 2.8 blocks.

"Jesus," I said, looking at the map. "We're going to the Back of the Yards."

"What?" Kyle asked. "Where's that?"

"Packingtown. It used to be the worst slum in the country."

The "yards" that the neighborhood was in back of were the Union Stock Yards, which for over a century had made Chicago into the world capital of death. More higher life-forms have perished in Chicago than in any other place on earth.

Imagine the globe rendered in some sort of death-sensitive imaging system (infradead photography?), whereby areas of increased mortality are highlighted or, perhaps more fittingly, darkened. Darkness rolls over the whole globe, of course, but Dachau, Buchenwald, Birkenau show as pinpricks of black, along with Hiroshima, Nagasaki, Chernobyl, Bhopal, the World Trade Center. But Chicago—Chicago of the area where Joe, Kyle, and I were headed, the district bounded by Pershing Avenue, Halsted Street, Forty-seventh Street, and Ashland Avenue—Chicago would be a massive black hole, a sucking vortex that swept billions of creatures into oblivion and onto dinner plates.

Animal rights activists refer to the place as the "Inferno." From the time the Union Stock Yards and Transit Company opened on Christmas Day 1865, through the development by Gustavus Swift of the first practical refrigerated railcar in 1882, until decentralization and interstate trucking doomed the enterprise in the years after World War II, Chicago killed the country's cattle, hogs, and sheep with a brisk and brutal midwestern efficiency.

Wags called it the "disassembly line." Decades before Detroit was a glimmer in Henry Ford's eye, first Cincinnati (aka "Porkopolis") and then Chicago perfected the assembly-line slaughter of animals. The cattle cars arrived by rail from all over the country, but mostly from the rangelands of the western states, and were shunted along the 130 miles of railroad tracks within the Yards. Animals for slaughter were first stunned with a poleax, and later, as techniques were refined, with pneumatic guns equipped with captive bolts. The actual killing occurred as their throats were cut and they bled out, "desanguinated," to use the terminology of forensics. Using power hoists and other innovations, workers hung the carcass overhead on a moving chain, passing it between cleaver-wielding cutters, each responsible for a different slice of meat.

"No iron cog-wheels could work with more regular motion," said visitor Frederick Law Olmsted of the butchers on the killing line.

The whole offal-and-manure mess of the yards drained into the South Fork of the Chicago River, which locals called "Bubbly Creek," since rotting animal carcasses rendered the water effervescent. The neighborhood had a lot of names. Ducktown, because it flooded so often. Canaryville, because of the sparrows that fed off the feedlots, and also for the gangs of young toughs ("wild canaries") that infested the place.

By the end of the nineteenth century, when the Yards cranked up to spectacle-level production, tourists flocked to the site. One of the recognized side attractions during the 1893 World's Columbian Exhibition was a day trip to the Union Stock Yards. The actress Sarah Bernhardt termed her visit "a horrible and magnificent spectacle," and Rudyard Kipling stated simply, "You will never forget the sight."

In 1906 Upton Sinclair portrayed the pathos of the place in *The Jungle:*

> Now and then a visitor wept, to be sure, but this slaughtering machine ran on, visitors or no visitors. It was like some horrible crime committed in a dungeon, all unseen and unheeded, buried out of sight and memory. One could not stand and watch very long without becoming philosophical, without beginning to deal with symbols and similes, and to hear the hogsqueal of the universe.

By 1900, the Chicago meatpackers represented a monopoly as complete as John D. Rockefeller's stranglehold on oil. The Union Stock Yards' 475 acres of holding pens, abattoirs, and rendering plants churned out eighty-two percent of the country's meat, as well as untold amounts of by-product-based items such as leather, glue, soap, oil, hairbrushes, fertilizer, faux ivory piano keys, oleomargarine, gelatin, shoe polish, buttons, perfume, and violin strings.

The far-thinking, refrigeration-minded Swift was joined by moguls Philip "King Meat" Armour, Edward Morris, and G. H. Hammond in shipping processed beef and pork not only to New

York and the other major metropolitan areas of America, but all over the globe, to London, Tokyo, Spain, Bordeaux, and St. Petersburg. Armour's motto* was "We Feed the World."

On the killing and cutting lines, fifty animals died every minute of every hour, twenty-four hours a day, 365 days a year. Death took no holidays. Conservatively totaling the numbers involved yields a ballpark figure worthy of U.S. Cellular Field: 2,210,000,000 animals perished over the course of the rough century that the Union Stock Yards were in operation.

Chicago's complex relationship with meat can be illustrated with an anecdote from the last Beatles concert tour of America in 1966. At shows around the country, fans pelted the band with jellybeans (a candy that John had once made the mistake of saying they enjoyed) as well as stuffed animals, lingerie, and love letters. But at the International Amphitheater in Chicago, on the site of the Union Stock Yards, a fan winged a frozen slab of sirloin steak at the stage, nearly beheading Paul McCartney. I can't be entirely certain, but I would venture to bet Chicago holds the distinction of being the only stop on the tour where meat was thrown.

Joe guided the white Aftermath truck south off I-55 at California Avenue and past the former site of the yards, now an industrial park. We plunged into a rind of a neighborhood, a *quartier perdu* that still exhibited its working-class heritage. The pubs and Catholic churches spoke of Poland or Ireland, but some of the stores were Mexican groceries. This was the fabled "Back of the Yards," formerly thick with Union Stock Yards employee immigrants and their families, a festering sinkhole so tubercular that at

---

* Not to be confused with the company's insidiously memorable hot-dog jingle, advertising "the dog kids love to bite!"

the turn of the 1900s, "newborns coughed blood coming out of the womb," according to reformer Mary McDowell.

Our destination was a four-story building with a stained, gray-stuccoed exterior. "A retirement home," Joe said, but of the single-occupant-apartment variety, designed for the not-quite-ready-for-the-wheelchair elderly.

Joe was nervous about his first body removal job, constantly on the cell to his mentor, Ryan O'Shea, for some hand holding. He had us gear up extra carefully: yellow heavy-duty hazmat suits over the more normal white ones, duct-taped at the arms and feet, full respirators. I felt fat.

"You look like a yellow snowman," I said to Joe, but the respirator dimmed my words, and he only smiled vaguely.

We waddled inside.

In a glass case next to the elevator, a computer printout with a pastel, flower-topped casket graphic announced the death of Alvin Kamolinski of Apartment 412. I could tell the graphic had been used before, and that the front office of the retirement home kept it at the ready, just switching the particulars as necessary.

Several residents quailed visibly as they passed by us waiting for the elevator. I was so caught up in the process of gearing up that I hadn't considered the effect the sight of us might have, and that those near death's door might not appreciate being ambushed by apocalyptic figures come to fetch one of their own.

A lone security guard wearing a blue nylon jacket a couple sizes too large waited for us outside the apartment on the fourth floor.

"In here," he called as soon as we left the elevator.

We were ambushed ourselves as we walked through the door of the apartment. The body lay facedown on the linoleum floor of the kitchen just to left of the entryway. It came as a shock to me (and, judging from a stutter in their movements, to Joe and Kyle too), even though I had known it would be there.

An in-house physician had already issued a death certificate, but there had been some delay getting the medical examiner's office to release the body. The postmortem interval, according to the retirement home management, had already been six hours. Which meant algor mortis, or body cooling, was over, and the body was in the midst of the extended seizing up known as rigor.

"We lift him up," I said to Joe, "he's going to be like concrete."

Kyle broke open a body bag. Aftermath normally used body bags (more politely called postmortem bags) that were made in China and manufactured out of three-mil polyethylene film, the same black plastic material used for trash bags, with a scrim of reinforcement mesh sandwiched between the double layers. But the company was trying out a newfangled gray thirty-six-by-ninety-inch TCP-CF3 Waffel Bag ("Proudly Made in the USA," read the packing slip) with an envelope-style #9 zipper for easy access.

"We get a lot of compliments on our body bags," Ryan O'Shea told me later. Not from their occupants, of course, but from police and coroner's personnel. They were a two-for-one deal: The clear polyethylene packaging that the postmortem bags came in was "Adaptable for Holding the Personal Belongings of the Deceased."

I knelt beside the deceased, a gaunt, sixtysomething gent with a smooth-shaved head showing a stubble of gray hair. Hair and fingernails do not really keep growing after death, as the common wisdom insists; it's only that rigor makes the skin recede somewhat at the follicles and cuticles, thus giving the impression of growth.

Statisically, most of us expire in the bathroom. But the end

had come to Alvin Kamolinski while he was in his kitchen in relative undress, bare to the waist, answering the boxers-or-briefs
question with the former, wearing brown vinyl slippers and spot-
less white crew sox. His shanks were reed thin, maculate with
liver spotting, but otherwise almost as pale as his hose.

"Where are the police?" Joe asked Jack, the security guard.

"They said they couldn't come right away, be maybe by tomorrow," Jack said.

"You're kidding," Joe said.

"They're busy," Jack said. "Must be a real busy day, lots of
crime or something."

"That sucks," Joe said.

"Mr. Anders said we have to get the body out," Jack said, not
bothering to explain who Mr. Anders might be. "It's disturbing
the other residents."

I stood up and looked across the room divider into the tiny
kitchenette apartment. Stacks of magazines, and an unused
treadmill hung with laundry. Model-train paraphernalia, much of
it still boxed, gestures toward a halfhearted hobby.

"Mr. Kamolinski always had a nice word," Jack said. "I never
like to see them this way."

"We can bag up the body," Joe said, "but I don't know if we
can transport it without the police."

"What do you mean?" Jack said. "I told you, police say they
can't come, and we got to get this out of here right now."

"That's the law," Joe said. "If we have this body in our truck,
we need to have a police escort, one car ahead, one car behind,
otherwise we'd be illegal."

Different jurisdictions have different rules and different levels
of respect for the dead. The Aftermath techs like to tell the leg-
end of Greg Banach's midnight ride, when he transported a

corpse from a job in his own SUV. The body was that of a tall man, and Banach crammed the corpse into the back of his truck, with its bottom half visible through the back window.

Jack was miffed. "Why'd we call you all, then? You said your company does body removal."

"You want us to do what we can do?" Joe said.

"What, put him in the bag and then just leave it here?" Jack said, shaking his head. I noticed the effortless transformation in his words. Mr. Kamolinski went from "him" to "it" in the space of a few words.

"I don't know what else we can do," Joe said.

"I got to talk to Mr. Anders," Jack said.

He opened the door and stepped into the hallway. We heard the squawk of his two-way. Kyle went with him, and Joe left for the truck to call Chris and Tim about the impasse.

"Hollywood can stay here," Joe said. He had picked up my handle from Ryan and Dave. "You don't mind, do you?"

I told him that I didn't mind. I was left alone with the body.

"How do you do it?"

That's a common question bystanders, friends, and perfect strangers put to the Aftermath techs, when the subject of their job comes up. What the question usually means is not "What procedures do you use to clean up crime scenes?"—although people want to know about that, too, down to what kind of chemicals are involved—but "How do you get used to it?"

It's a philosophical query, posed by the ninety-one-point-five percent of the population who don't deal with death on a daily basis to the eight-point-five percent who do. The question implies that continual exposure must somehow transform the techs, render them abnormal, compromise them in some fundamental way. Butchers in Elizabethan England were not allowed to serve on juries because, the reasoning went, their close familiarity with death desensitized them to human suffering.

With a gloved hand I felt the back of Mr. Kamolinski's neck. I tried to reposition him to see his face, but rigor had temporarily locked him in its embrace. The blotchy stains of livor mortis mottled his skin. I touched the pressed-wood kitchen cabinet next to his head. Human body and wood cabinet were the same temperature, and gave off the same sense of physical inertness.

Cool flesh. In Switzerland, banks warm their marble counter-tops to skin temperature, which supposedly makes money easier to pick up. The normal temperature of human skin is 91°F, or 33°C. The ambient temperature that fall day was closer to the mid-seventies, even without air conditioning, so within the last six hours, the body's skin temperature had dropped by 16°F. Living cells are microscopic furnaces, fueled by sugars, fed by oxygen. Algor mortis is what happens when the furnace goes out. An energy exchange occurs, with the dead body giving up its heat to the surroundings, aligning itself with the ambient state of the universe. From he to it. Becoming a thing.

How do we do it? How do we become jaded, inured, accustomed to death?

....

I had seen Mr. Kamolinski before. Not him exactly, of course, but someone who looked a lot like him and, more important, had shuffled off the coil the same as he had. The first time I laid eyes on a dead body might have been poor, unfortunate Chucky Sipple in Wisconsin, but the first time I ever touched one was in the early 1980s, at a school in the Medical District near the University of Chicago, only a few miles from where Mr. Kamolinski was cooling his heels.

My girlfriend at that time was studying to become a doctor, and she took me to the medical school's gross anatomy lab* after hours, when no one was around. Strictly forbidden, of course, and she could have gotten into serious trouble had we been caught, but Tina Fishman had been born a youngest child, and had a classic youngest-child rebellious streak.

The laboratory was dark when we entered. Tina switched on the banks of fluorescent lights on the ceiling. Two dozen embalmed, partially dissected cadavers lay on stainless steel gurneys in six rows of four. Some of the bodies were covered in white cotton shrouds, but others had been casually left partially exposed.

I turned my head away and retched.

"You okay?" Tina said. "Can you handle it?"

"Sure," I said, determined to soldier on. *Nothing human is foreign to me, nothing human is foreign to me, nothing human is . . . ugh.*

"We each get our own," Tina explained. "Here's mine."

She pulled back a sheet to reveal an elderly, beak-nosed Caucasian male with a shaved head. An empty body cavity indicated where his viscera should have been, with the whole surprisingly compact lungs-heart-liver-entrails package removed and set down between his legs. Working in class, Tina had also partially peeled away one side of her specimen's facial skin.

"Some people donate their own bodies, but most of them are like homeless guys, destitute. We get them, and if there is any

---

* *Gross* used in this sense is not a value judgment, but invokes "total" as in the phrase *gross income.* Although the lab was gross ("vulgar, obscene, coarse") also.

family, it gets, you know, some money to pay a funeral home for a memorial service or something."

The balking chemical stench of formaldehyde masked any scent of decay. I bent down and stared at the cadaver's one good eye, its pupil occluded with a milky haze. I placed my ungloved hand on its forehead. Clammy.

"Bobby," Tina said. "That's what I named him."

"Was his head shaved like that or did you do that to him?"

"I did it with a scalpel," she said. "He had some wispy stuff, like tufts, you know? But I didn't like the way it looked, so I cut it off."

"Jesus," I said.

"What?"

"You make it sound like you own him."

"Well, he is mine." She petted the head of her cadaver, also with an ungloved hand. "My Bobby," she said.

She asked me if I wanted to see where the cadavers at the lab came from and I said that I did. We drove along Harrison Street parallel to the Eisenhower Expressway, to a nondescript brick building on the grounds of Rush University Medical Center.

Tina knew the young Hispanic caretaker of the place, and introduced me to him as "an undercover reporter." That meant essentially nothing, but the guard accepted it without question. In the basement of the building was a large swimming pool, formerly used for sports and recreation. Only, this pool was filled with formaldehyde, and a dozen blank-eyed, gray-skinned cadavers floated in it, men and women, black and white, but uniformly aged.

The chemical smell was overpowering. The guard poked and pushed the corpses beneath the liquid, using a proverbial ten-foot pole of aluminum. "The guy before me," he said, "had to quit because he got allergic. But it don't bother me."

I couldn't take it very long. I felt like *I* was becoming allergic. "What'd you think?" Tina asked, as we exited into the fresh midwestern air.

I hesitated before answering. "There's this fear of being just a body, you know? Like we're just meat," I said. "Don't you feel that?"

That was the mirror that Aftermath held up to me, again and again. The mind-body duality finally resolved, and human life reduced to mere flyspecked meat, to amber fluids and black stains. A lot of people (over ninety percent of the population, it turns out) spend their lives assiduously avoiding looking into that mirror. Religion is all about that avoidance, and romanticism, and hedonism, and probably all the other isms the flesh is heir to. We pin all our hopes on transcendence. Nobody, least of all Mr. Kamolinski, wants to be just a cold slab of meat left alone on the floor of a kitchenette.

I hunkered down to keep him company.

. . . .

Chicago, the city of death, did not give up its dead easily. The Back of the Yards job was an exception. Aftermath, Inc., had yet to crack the Cook County market. The company was largely a suburban, not an urban, concern.

"We don't get that many referrals within the city of Chicago," Tim told me. "It's all pretty tightly controlled." Chris was less guarded. "It's political, it has to be—they always give their cleanup jobs to their own buddies."

Before the police-escort impasse, Joe and Kyle had been directed to remove Mr. Kamolinski's body to the Cook County Medical Examiner's Office, near the United Center, where the

Chicago Bulls play. Joe got on the phone with Ryan, considering how to transport the body.

"Do we put it in the back of the truck and just bungee it in place?" Joe asked.

He listened, then shook his head, repeating what he heard to Kyle. "Ryan says we should put it up front in the cab, just prop it up there between us."

"Hey, that's where I sit," I said.

"I think he was kidding," Joe said.

It would have been a short trip straight north, but still, none of us relished a journey in the company of a corpse.

The Medical Examiner's Office was on Harrison Street near the University of Chicago hospitals, just down the street from where I first encountered Tina Fishman's Bobby. A few weeks before the Back of the Yards retirement-home job, I had called the office to request an informational tour. I discovered that journalists were not at all encouraged to investigate conditions at the Cook County morgue.

"We give tours to police officers, and that's it," a morgue functionary told me over the phone. "I've never heard of anyone else having access, and I'm the one who sets up the tours."

He suggested that I might query the medical examiner of Cook County, Edmund Donoghue, directly. I faxed over my request, and got a call from Donoghue a couple hours later.

"What you propose just isn't going to happen," Donoghue told me. Donoghue was a surgeon's son, a Chicago boy born and bred. "It's a conflict of interest," he said.

"I don't really see that," I said.

"This Aftermath, it's a business, and we can't show any preferences like that."

In my faxed letter to Donoghue, I had mentioned that I was

making my request "in order to gain a better understanding of the activities of a major urban medical examiner's office."

"If you want to see what a big-city morgue does, you can do it out where you're from, in New York," Donoghue said. I caught a tinge of the Second City's New York phobia in his tone.

This wasn't going the way I had hoped. I was running up against either Cook County's traditional wall of silence or the larger, equally traditional pattern of shame about the mechanics of death. I considered mentioning sunshine laws, transparent government, media access to truth.

"Well," I managed, "perhaps you and I could sit down for an interview, without the tour."

"This isn't a negotiation," Donoghue snapped, "this is a no."

"Wow," I said, unable to help myself. "You really need some media training. I know professionals who can help in that area—"

But Donoghue had already hung up.

Later on, about a week before the Back of the Yards job, I managed a sub-rosa visit to the Cook County morgue. My guide, who I will refer to only as "Mrs. O'Leary," cheerfully conducted me through the whole of the premises, including a peek into the private offices of the medical examiner himself, the absent Donoghue.

I was more interested in the morgue proper. Off a well-lit hallway, three doorways led into a series of holding and autopsy rooms. The first room we entered was a large cold-storage facility, measuring forty feet by sixty, with twelve-foot-high ceilings. The air here was kept at around forty degrees—not cold enough to see your breath, but cold enough to need a jacket.

Along the walls were shroud-covered bodies (I counted fourteen) resting on metal gurneys that were slid into wooden-framed racks. For the entire length of the room the wooden shelving units stacked three levels high. Some of the slots were empty, and the rest seemed occupied randomly, with no discernible pattern. The final effect was pigeonholes indifferently filled.

"These are mostly unclaimed," my guide said. "We're waiting for families to come for them or to be contacted."

"What if there are no families?"

"They'll have a state burial," she said. She stared around at the supine, sheeted figures punctuating the walls. "Some of these have been here for coming up on two weeks."

One of the bodies attracted my attention because its sheet poked upward from underneath very awkwardly. My guide lifted the sheet to reveal an elderly male with a grizzled face and a deep crater where the back half of his skull should have been. The male's arms were outstretched in a cocked, supplicating posture, as if he were pleading for grace.

"That's not rigor," my guide said quickly, dropping the sheet. "He was homeless, sleeping outside in the cold. He was murdered; his killer bashed the back of his head in with a rock when he was asleep. The temperature the night we found him was in the low teens, and he had frozen in that position. They brought him in here, and he still hasn't thawed."

Some die by ice, and some by fire. The bodies of two Chicago firefighters, killed when a floor collapsed under them, exhibited the full pain and trauma of their deaths. The disfigured corpses offered graphic evidence that extreme heat, when applied to human skin, brings out all the similarities between human flesh and pork. I felt revulsion and compassion at once. The skin of their faces was "alligatored," shriveled by the flames into a patchwork steakhouse char.

The next room was smaller and not refrigerated. Two stainless steel autopsy tables stood side by side, both eight and a half feet long and two and a half feet wide. Autopsy tables, like subway seats and school desks, have had to be enlarged over the past few decades to accommodate the increasing girth of the population.

Generally, any surface that dead flesh touched within the morgue was stainless steel. Each table sported its own scale and a sink at one end. Goosenecked hoses rose from beneath, with convenient knee-operated hot and cold water controls. I wondered if the waste from Donoghue's autopsies discharged into the communal sewage conduits. Was it treated first? Filtered? At the big water treatment plant in Skokie, did the gore mix promiscuously with sewage? Was it a case of shit *and* death?

There were bodies in this smaller room, too, not as many, under a half dozen, readied for cleaning, processing, and storage. Through a glass-windowed wall, I had seen a lab-coated forensic pathologist with an examination under way. Not an autopsy, at least not yet. But the tables in this main room were empty, glistening and receptive. I was fascinated by the accoutrements of the tables, used to prop the bodies up during an autopsy: stainless steel headrests, rubber neck supports, longitudinal body frames.

I hopped up and lay down on the nearest table, eliciting a nervous laugh from Mrs. O'Leary. "That's probably not a good idea," she said.

I told her that Sarah Bernhardt used to bring along a casket on her tours, and sleep in it backstage. I imagined my own guts spilled onto the autopsy table, poked and prodded by Edmund Donoghue.

The word *coroner* derives from the phrase *custos placitorum coronae,* "guardian of the rights of the crown," and originally referred to an officer of the royal household in medieval England. When the various crows picked over the bones of the dead, the coroner was there to make sure the crown got its share. Suicides forfeited their possessions to the crown, so the coroner got involved in cause-of-death judgments.

In England and Wales, a coroner must be a solicitor or barrister, but as the function evolved in the United States, it became an elective office open to anyone, and as such famously corrupt.

The trend toward medical examiners in the last century represented an attempt to professionalize the office, and remove it from the possibility of political suasion and malfeasance. That's probably what Donoghue thought he was guarding against, when he mentioned "conflict of interest." In Cook County, historically the site of so much official larceny, the ME had to be like Caesar's wife, above reproach.

"This isn't the only autopsy room," Mrs. O'Leary said, entering into her full tour-guide mode as I rose, Lazarus-like, off the table. "We handle four thousand five hundred autopsies every year, out of twelve thousand deaths that are referred to the ME. There are over a dozen pathologists and a whole crew of investigators."

Next door a smaller, colder cold-storage room with its own autopsy table held the dozen remains of unattended deaths, bodies in various degrees of decomposition. This room was seriously chilled, its temperature hovering just above freezing, in an effort to quell the smell. My gorge rose. The stench of decay was stronger here, and I soon found out why.

"There's one in here three, four weeks decomposed, and you can really smell it," Mrs. O'Leary said.

She pulled the sheet off a decayed, partially mummified corpse that showed signs of both animal predation and the development of "grave wax," adipocere, which meant the body had been exposed to damp conditions. I stared into the mottled, grease-caked visage, trying without much success to stir my own sense of empathy.

Exactly here was the genesis and foundation of the idea of power, according to Mahfouz: a live human being looking at a dead one. From that age-old, elemental tableau spun the webs of authority, supremacy, and control in which society was caught up. Looking at the face of death, the living human exulted, feeling a surge of omnipotence. Like passing a hapless pedestrian when driving in a car.

My own response turned out to be less perspicacious than that of the great Egyptian Nobel laureate, but it amounted to essentially the same thing. Underneath my anxiety about getting busted for trespassing in the Chicago morgue and my worried, busybody assessments about what kind of notes I needed to take, I detected a small thrumming interior voice repeating the same two words over and over.

Not me.

Or actually the phrase deserved an exclamation point. *Not me!* Maybe also caps. *NOT ME!*

I discovered that my pity for the dead was always underpinned by a craven sense of relish. Beneath the proper, culturally acceptable sadness lurked flashes of dimly lit emotions, furtive and unexposed, like far-off lightning punctuating a nighttime sky.

Relief, yes, and what the Germans call Schadenfreude, glee over another's misfortune, and a coldhearted selfishness. As I brought my face close to the anonymous grave-waxed corpse I felt myself gripped not by grief but by mundane personal worries: that I would contract a disease, that I felt discomfort in the refrigerated cold, that I might be plagued at night by bad dreams. Bending a few inches from the dead man, "not me" morphed imperceptibly into "me, me, me!" My temporal concerns overshadowed his eternal ones. My hangnail, your cancer. A fundamental human equation.

How do we get used to it?

"Cast a cold eye on life, on death," runs Yeats's epitaph. I knew there existed among us people who had mastered the cold-eyed approach to mortality and fellow-suffering. Chicago, more than any other place in America, spawned some notorious examples.

....

On the evening of June 13, 1982, a seventeen-year-old African-American street prostitute needed a date to get off the street. The other girls on Chicago's North Side trolling grounds called the seventeen-year-old "St. Louis," because she'd just arrived from the East St. Louis whorehouse town of Brooklyn, across the river from the Gateway City.

"The police were hot that night," she testified later in court. "They were picking up all the girls."

She reacted with relief when a red Dodge Tradesman van bearing the logo "R & R Electric" cruised up next to her on Elton Avenue. Because of all the police heat, St. Louis would have gone along willingly with the shaggy-haired thirty-year-old driver, an electrical contractor and handyman named Robin Gecht. Instead, Gecht pulled a handgun and forced her into the back of the van.

Gecht bound her waist with a thin metal wire, fastening the other end to the aluminum shelving that lined both sides of the van's interior. St. Louis remembered seeing a lot of evil-looking electrician's tools on the shelves. Gecht raped her, but did not untie her when he finished.

"Don't kill me," St. Louis pleaded. "I'll do anything you want me to."

Still training his pistol on her, Gecht placed a six-inch hunting knife in her hand and told her to use it on herself.

"Stab into your tit," he said. "The left one."

Gecht dragged a white plastic bucket over and propped it under St. Louis as she did as she was told. The blood from the wound drained into the bucket. Gecht appeared transfixed, not by the blood but by the punctured breast. He poked his hand at the inch-long cut, tugging open the skin, first with his finger, then with a butcher knife. The mutilation excited him, and he sexually assaulted St. Louis once more.

St. Louis was one of the few lucky ones; she lived. That summer and fall, the electrician's van became the object of street rumor among the working girls along North and Elton Avenues. *If you see the red van, don't get in.*

Much deeper wounds could be inflicted if a young woman fell into the hands of Robin Gecht and his coterie of young employees. The tabloid *Chicago Sun-Times* would label the team killers "the Chicago Rippers," for their Jack the Ripper–style mutilation murders of more than a dozen young women.

In the attic of Gecht's rented two-story brick home at 2163 North McVicker Avenue, in a working-class neighborhood of the near North Side, a closet-sized room bore crudely painted black and red crosses on its walls. Baby-sitters for Gecht's three young children were warned never to venture upstairs to the attic. When one did, she found a doorway heavily barricaded with plywood.

Tommy Kokoraleis, a twenty-two-year-old who played Renfield to Robin Gecht's Dracula, told police what happened in that attic room. As Gecht's mutilation fetish became more and more unmanageable, he sent the feebleminded Tommy, Tommy's brother Andy, and their co-worker Eddie Spreitzer to kidnap and kill victims. It is difficult to convince others to murder in your name, but Gecht was a rare, Manson-like figure who managed it, mainly by employing rough, impressionable street-kid males at his electrical contracting business.

With or without Gecht aboard in the Dodge Tradesman, the others trolled for street prostitutes, lone women, runaways. Gecht always specified large breasts. (In a letter to a journalist, Gecht would write from prison that well-endowed women "were a thing with my whole family," and praised his wife Rosemary's 38D bra size.)

Over the fall of 1981, into the summer of 1982, the attacks and mutilations became more and more extreme. Gecht had his cohorts would kidnap young women, rape them, drug them. They would often garrote one of their victim's breasts with thin copper electrician's wire, tightening the wire until it severed the breast completely. Tommy Kokoraleis testified that Gecht would sometimes "have sex" with the wounds, and with the severed breast.

Afterward, the gang of four would repair with their trophies to Gecht's attic room with the red and black crosses on its walls. At a makeshift altar, in a perversion of the Eucharistic rites, each of the Rippers would eat a piece of "breast meat," as Gecht called it, after which their leader would carefully preserve the remainder in a small decorated box.

I knew the story of the Rippers well. During the period when I spent time in Chicago with Tina Fishman and met her Bobby, local newspapers and newscasts were filled with lurid tales of Robin Gecht and his friends. I read the stories with a mix of fascination and stupefaction. This was also during the time of Tylenol tampering, with seven Chicagoans dead from cyanide poisoning, when it seemed as though the city, never a law-abiding paradigm to begin with, was going to wobble completely off its axis.

"Suspect linked to 'cannibal gang' gets life," read the *Chicago*

Tribune on September 8, 1984, after Judge John Nelligan of Du-
Page County Circuit Court sentenced Thomas Kokoraleis. The
Ripper killings came hard after the homosexual-panic murders of
John Wayne Gacy, a decade earlier, in the same Chicago neighborhood. There was the unsettling news that Robin Gecht had
once worked for John Wayne Gacy. Had he apprenticed for
something more than electrical contracting? Could serial killing
be a virus, transmittable from one person to another?

The list of toddlin' town monsters was actually pretty long. Dr.
H. H. Holmes followed Jack the Ripper by only a few years, and
killed most of his victims in Chicago. Another "doctor," though
not a medical one, Theodore Kaczynski, the Unabomber, also
hailed from Chicago, as did the nurse-killer Richard Speck.

"The long pig" translates a Maori word, *pakeha,* used by Mi-
cronesian cannibals to describe humans as food. Man as meat.
Under the bad-moon influence of the Union Stock Yards ("over
two billion served!"), and strengthened by stories of the Rippers,
the long pig had become associated with Chicago in my mind
long before I signed on to Aftermath.

I observed it in the obsessive nomenclature of area rock
bands: Cannibal Cheerleaders on Crack, Cannibal Corpse, Cannibal Ox, Cannibal Galaxy. The turf of Jeffrey Dahmer, the most
infamous flesh-eating serial murderer of them all, was but a
short Wisconsin death trip to the north, in Milwaukee.

I didn't want to believe that I had anything in common with
the whole cold-eyed lot of them—Dahmer, Gacy, the Rippers,
the long parade of sociopaths that stalked the land where I was
born. But looking at the rotted cadaver in Edmund Donoghue's
keep, or staring down at the lifeless Mr. Kamolinski, I found that
I could locate in myself the same immoral, amoral attitude that

transformed flesh into meat. The attitude of a Carib, gazing appraisingly at a long pig.

Did working at Aftermath encourage that attitude? Or lessen it? I had now been job-shadowing techs off and on for three months. Was I being transformed? I couldn't decide for sure. No Jung-Myers-Briggs personality test existed to gauge my increasing (or decreasing) heartlessness. Would I finally become bored with the puzzle of mortality, which was, after all, insoluble? I worried that the more quotidian it became, the more death would be stripped of its mystery. And I couldn't decide whether that was a good thing or not.

As a kiss-off gift when we broke up, Tina Fishman sent me Bobby's penis. She boxed and shipped her cadaver's decollated sexual organ, glans to testicles, to my work address. When I opened the package, I found that Tina had carefully tucked the gray, shriveled penis into a hot dog bun and scattered ketchup and mustard packets around it. ("What kind of kid likes Armour hot dogs? Big kids, little kids, kids who climb on rocks! Fat kids, skinny kids, even kids with chicken pox like hot dogs!")

The message, as I understood it back then, was something along the lines of "You're a weenie," a not-uncommon sentiment in breakups. But I had never encountered it expressed in quite so forceful a way. I tossed the dingus in the trash, bun, condiments, and all. Later on, I had second thoughts. Maybe Tina hadn't really meant to insult me. She knew that I was just perverse enough to appreciate such a macabre gift. What I should have done, I considered afterward, was buy some formaldehyde, discard the bun, and keep the dog, showcasing it in a specimen jar on the mantel. A conversation piece.

Even later in retrospect, I had third, fourth, and fifth thoughts.

Word of Tina's odd gifting got out in gossip around her medical school. She had to sit before a medical ethics panel, which briefly suspended her but decided against expulsion. The authorities, who might have charged Tina with "unlawful disposition of human remains," were not notified. She went on to become a pediatrician in Vermont, her body-robbing days long past.

The long pig pokes its snout everywhere, into individual lives and into odd historical crannies. Henry Ford learned everything he knew about the assembly line from the Union Stock Yards.

"I believe that this was the first moving line ever installed," Ford writes of the Chicago Yards in his autobiography. "The idea [for the Ford Company's Detroit assembly line] came in a general way from the overhead trolley that the Chicago packers use in dressing beef."

The anti-Semitic Ford died just as revelations about the German extermination camps were coming to light, but he might have been heartened to know that his assembly-line ideas had been adopted by the Nazis to implement the Final Solution. An aerial view of the 425-acre Birkenau, the death camp at Auschwitz, with its branching lines of railroad tracks, bears an uncanny resemblance to the 475-acre Union Stock Yards. In both places, victims were delivered up in cattle cars.

"King Meat" Armour and Gustavus Swift to Henry Ford and Alfred Sloan, and finally to Heinrich Himmler and Josef Mengele. Hogs to gas hogs to the long pig. Tinker to Evers to Chance—Chicago Cubs all. Connect the dots.

Numerous historical examples exist of pigs being prosecuted for the murder of humans. Hauling animals into court represented a common enough practice in the Middle Ages. In 1386, for example, a sow was convicted of killing a child. Authorities dressed the swine culprit in men's clothing and held a public execution. But as a rule the reverse is never true. Humans are not criminally prosecuted for killing hogs.

At the age of two, Mozart identified a pig's squeal as being a G-sharp. So the Chicago Yards and Hitchcock's *Psycho* shower scene pig squeal sounded the same note. Illinois has an official state bird (the cardinal), and an official state flower (the dooryard violet). If the city of Chicago ever decided on an official musical note, it would have to be G-sharp.

At the Back of the Yards retirement home that day, we never secured our police escort. Joe, Kyle, and I left the corpus of Mr. Kamolinski bagged in his kitchenette, to be retrieved whenever Edmund Donoghue and the Cook County authorities got around to getting it. We drove the Ronald Reagan Memorial Tollway back to what I was rapidly coming to consider, with fondness and loathing, as "Aftermath country," where the keepers of the long pig were either more accommodating or had less to hide than they did in the city of Chicago, and the hogsqueal of the universe did not sound quite so loud.

INTERLUDE TWO:

Nonlethals

A meth lab burning

One has a right not to be fallen on.—John Berryman

Look at me, busy as a bee! Where'd I get all this
energy? Whoa-oh, meth, meth! I don't sleep,
and I don't eat. But I've got the cleanest house
on the street!—Song lyrics in the public service ad "Cleaner
Girl," National Youth Anti-Drug Media Campaign

You couldn't ask for a more peaceful, bucolic scene that Thursday afternoon, August 25, 2005, on the back deck of Sarah Searer's home in Villa Park, suburban Chicago. The white two-story house on a cul-de-sac overlooked the meandering Salt Creek, just north of Salt Creek State Park. Sarah relaxed after lunch, with a Bible and a copy of *Today's Christian Woman* magazine beside her.

It was just after 1:00 P.M. Sarah's fifteen-year-old son, Dustin, mowed the grass. Her other son, Ben, nineteen, sat inside the house at the computer.

Sarah looked up to see a stocky DuPage County sheriff's deputy splashing full-tilt across the creek toward her property. The uniformed deputy brandished a 12-gauge pump-action shotgun. For a split second the incongruity of the scene froze her in place. Then she acted.

Sarah opened the sliding glass doors on the deck, grabbed Dustin, and rushed with him into the house to gather up Ben.

Rosie, the family's lab-chow mix, barked furiously toward the top of the stairs. Then another incongruous image struck Sarah with new terror: A slender Hispanic male stood at the top of the stairs with a small silver automatic in his hand.

Sarah and Dustin retreated quickly out of the house the way they had come in, cutting through the side yard and reaching the safety of the street.

Ben was caught.

The intruder had slipped in the open front door and locked it behind him. In his panic, Ben had difficulty with the lock. Rosie barked and snarled.

With the intruder only a few feet away on the stairs, Ben heard the sharp crack of the pistol. Rosie abruptly fell silent. The home invader had shot the dog.

Ben figured out the lock and tumbled outside. The Searer family was safe. Minus a dog, but safe.

Their home, however, became the site of a twenty-seven-hour standoff between the intruder and legions of police. Using percussion grenades and twenty-seven canisters of pepper gas, DuPage County sheriff's deputies, a SWAT team, and agents of the FBI performed a sustained assault on the lone dog-killing gunman holed up on the second floor of the Searer house.

When Dave Creager and Ryan O'Shea inherited the scene a week later, a toxic chemical irritant—a residue of the pepper gas—coated every interior surface. The percussion grenades had blown out every window, but that didn't make it any easier to breathe inside the house.

"The place looked like a bomb hit it, because a bomb did hit it," Ryan said. "Lots of bombs."

"I think the police went a little crazy," Dave said.

The scene at the Searer residence in Villa Park points at a home truth in Aftermath work: Some of the most difficult jobs do not involve human fatality at all. "Nonlethals," as the techs refer to them, span a full spectrum of circumstance. At the simplest end, nonlethals include accidents where blood may have been spilled, but no one died. A shopper collapsed at a big home supply store, splitting his forehead open on a wooden floor pallet. The small bloodstain had been cordoned off with orange security gates when Ryan and Dave got to it, and required less than a half hour to clean up.

But high-end nonlethals can be some of the most complex and time-consuming Aftermath jobs of all. The Villa Park cleanup took Ryan and Dave five days and ninety man hours to accomplish. It proved an agonizing job. The police had dumped a full riot's worth of tear gas into a single two-story residence.

"The stuff activates in water," Ryan said. "So we'd work all day, go home and take a shower . . . yow! You'd start hurting all over again."

At the start of the week-long job, they could remain in the house for only a half hour at a time. By the end of that thirty-minute period, even using respirators, they'd be wheezing, their eyes streaming with tears. Every cloth-based object in the Searer house, most of the furniture, carpeting, and clothing, had been thoroughly contaminated and had to be thrown out. Ryan and Dave arduously cleaned every square inch of ceiling, wall, and exposed floor. They removed the bloodstain on the stairs, where Rosie had fallen.

The gas grenades employed by the DuPage Sheriff's Department were so-called triple threat canisters, which used OC pepper combined with CS tear gas aerosols. *OC* stands for oleoresin capsicum, with the same active capsaicin ingredient that renders chilis hot. While the pepper-based substance was an inflammatory, the CS (ortho-chlorobenzylidene-malononitrile) was an irritant, providing a one-two punch designed to thoroughly incapacitate a target—swelling shut the eyes, locking in the CS. (The third element in the triple threat was the nitrogen-propelled grenade itself.)

Although tear gas has been outlawed for use in war, police departments around the world use it for crowd suppression and riot control. CN (chloroacetophenone, with the commercial brand name of Mace) has given way to a much wider use of CS, named after the initial letters of the names of its inventors, Ben Carson and Roger Staughton. Interestingly, before a police department in the UK can be licensed to use CS as a personal incapacitant spray, or PIS, its constables must themselves be sprayed with the stuff, in order for them to know how a PIS feels.

Ryan and Dave got the story of what happened that August afternoon at the Searer house from several sources, first and foremost the family, then the news media and police. At 12:30 that day, a thirty-year-old ex-Marine named Juan Silvas entered Harris Bank in La Grange, a Chicago suburb six miles to the southeast of Villa Park. He carried a silver handgun and an assault rifle, but botched the robbery, escaping with only a thousand dollars in cash and the police closing quickly in behind him as he fled west.

Since the bank was in their jurisdiction, DuPage County Sheriff's Department deputies picked up the chase. Firing on his pursuers with his assault rifle, Silvas put a few holes in squad cars and shot out the windshield of one, wounding a deputy with flying glass. Racing north on I-294 in his 1996 Cadillac Deville, he blew through the toll plaza at Oak Brook, finally leaving the highway at Exit 13, in Elmhurst.

With police converging on him from all directions, Silvas abandoned his car and fled on foot into the woods lining Salt Creek. Finding the Searers' door open, he slipped inside. Immediately as the Searer family fled, the police arrived. A force of more than seventy-five sheriff's deputies, FBI agents (because of the bank robbery connection), and other law enforcement personnel cordoned off the dead-end street and surrounded the house.

Dogs don't have the tear ducts to react to the lachrymatory effects of tear gas, and thus if Rosie had lived, she would not have suffered in the gas grenade barrage that hit the house in repeated waves, from Thursday evening through Friday morning. Police entered the house Friday afternoon to find Silvas dead from a self-inflicted gunshot wound.

The next day the Searer family, still not able to enter the house for periods longer than a few minutes, buried Rosie in the backyard.

Juan Silvas had no wife or children, no debts, and no criminal record. He had worked as a cook at Camp Lejeune during his stint in the Marine Corps. He had five sisters and brothers, and they were baffled as to why he walked into Harris Bank that day. Nothing in his past presaged such a development.

After the Aftermath techs thoroughly remediated the site, the Searers moved back in with no ill effects. When she threw a homecoming barbecue, Sarah Searer invited Ryan O'Shea and Dave Creager to attend.

.....

Around the Aftermath office the most dangerous, reviled, and unwelcome nonlethals were meth lab jobs, which resembled Superfund cleanups in miniature, with the added risk of fire and explosion thrown in.

"I'd rather do a six-week decomp than a meth lab any day," Greg Banach told me. "Too many fucking things can go wrong."

In a sense, home methamphetamine production pays tribute to the great American tradition of free enterprise, "except it's for morons," as Banach said. Meth is the only psychoactive drug that can be easily and cheaply manufactured by someone without a degree in chemistry. With the facilities of an ordinary home kitchen, a single meth cook can produce enough of the substance to supply twenty users. Meth democratizes drug production. Recipes are widely available on the Web. Raw materials can be purchased at Wal-Mart. Anyone can do it.

A lot of people did. Meth use is actually on the decline nationally, but that fact provides little relief to areas hard hit by the blight. Because the smell of meth production is terrific, resembling the odor of an enormous well-fouled litter box,* rural areas with a lot of wide-open spaces are especially susceptible to the scourge. Aftermath worked meth lab cleanups primarily in the downstate counties of Illinois, in communities along the Mississippi River, or on remote farms in Iowa and Indiana.

Aftermath techs traded stories of the bizarre behavior of meth addicts. Meth stimulates neurochemicals such as dopamine that signal alertness, pleasure, or euphoria, and causes many users to lose grip on their minds. Transported by the drug's pseudosexual rush, groups of meth users indulge in orgies in front of their children, who are often left to scavenge food for themselves as multiday meth binges consume the lives of their parents and caregivers. The refrigerator in a meth house is always either empty or full of chemicals. Meth addicts ("tweakers") tend to fall off the face of the civilized earth, neglecting normal human conventions such as cleanliness, sanity, and gainful employment in favor of a monomaniacal devotion to the drug.

Probably behind only heroin, meth has acted prominently upon the world stage. It should come as no surprise, given the in-

---

* Meth cooks sometimes filter by-product gases from their labs through kitty litter.

famous frothing-at-the-mouth performances at Nuremberg ral-
lies, that Hitler's private physician, Theodor Morell, regularly injected the leader with methamphetamine. Likewise, John F.
Kennedy mixed steroids (a treatment for his Addison's disease)
with injections of amphetamine. More recently, North Korea re-
plenished its paltry cash reserves by manufacturing and supply-
ing meth to black markets in Asia and Australia.

The drug's midwestern arena is smaller. A pair of meth users
dialed a 911 operator during an Iowa blizzard, giving their location as nearby an apartment building in the area. From the
source of the call, the 911 dispatcher realized this could not be
possible, and tried to elicit the correct information out of the
meth-addled callers. Describe your surroundings, she suggested.
The male caller said he was looking at a frozen pond. With
groups of people crossing it. Chinese people. Carrying auto
parts, which, the caller said, the Chinese people were placing in
the branches of trees.

The hallucinatory callers, a couple named Michael Wamsley
and Janelle Hornickel, both twenty, were found after the blizzard
had run its course, frozen to death. Authorities discovered them
in a wooded area along the Mississippi River, far from the apartment building that they had repeatedly given as their location.

Aftermath techs were generally called in not to clean up the
body fluids of dead meth addicts, but to remediate their labs.
These were not the "superlabs" capable of producing hundred-
pound batches of crystal meth, but smaller house- or apartment-
based operations. Drug enforcement agents nickname them
"Beavis and Butthead" labs, after the demotic cartoon characters.

In fact, the word *lab* only charitably applies to the chaotic
jumbles of cook pots, plastic tubing, bottles, and tubs used in
homemade methamphetamine production, and often lost amid

the general clutter and filth of the tweaker lifestyle. Of the thirty-two chemicals involved in meth production, a dozen are toxic, and many are corrosive, explosive, or flammable.

During the 1980s, when meth production remained largely in the hands of motorcycle gangs, phenylacetone was the reagent of choice, but in the areas of the Midwest serviced by After-math, tweakers tended to steal quantities of explosive ammonia fertilizer from farmers and treat it with metallic salts, a highly unstable process. White gas for camp stoves can be another com-bustible ingredient, along with lye, iodine crystals, hidriotic acid, and phosphorus—the red material at the end of matches.

Meth cooks who use a common recipe send their acolytes out on "smurfing" runs, mad invasions of convenience stores or gro-ceries with the purpose of grabbing all the cold medicine avail-able. The cooks then refine the pseudoephedrine and ephedrine in the cold pills to make meth. Left behind for Aftermath are flammable solvents, chlorinated solvents, and acid bases used in the process. Cleanup standards permit no more than .5 micro-grams (one-half millionth of a gram) of detectable meth residue per square foot.

Ryan and Dave answered a meth call in southern Illinois, in a set of low-slung apartments, formerly a motel, outside of Cham-paign. When they arrived, they found a smoking char that clearly would yield nothing to clean up. A woman had moved into the complex, set up a lab, and promptly burned herself and a dozen other residents out. That's how a fifth of meth labs are found—when they explode or catch fire during the volatile process of making the drug.

In the middle of the burned-out six-unit apartment building stood the local fire chief, looking lost. Ryan and Dave briefly picked through the rubble with him. The apartments had been burned flat. The chemically fed flames burned so hot the aluminum window frames melted into pools of liquid metal.

"What do you think, guys?" the fire chief asked them, turning over a melted lozenge of burned plastic. "Do you think that this could be part of the lab?"

Ryan and Dave exchanged looks. "I couldn't believe it," Ryan said later. "Here's the arson investigator asking us, like we're the experts. He seemed, like, stunned."

"I don't think we can help you," Dave told him.

"He didn't know where to begin," Ryan recalled.

The clothes of the female meth cook had been set on fire by a small explosion atop a Bunsen burner. There were other tweakers present who, instead of calling 911, simply watched her burn until it was too late, and she had set the building on fire. Then the solvents used in the cooking process took over, and the whole place went up.

"She's up in the burn unit at Cook County Medical." The fire chief looked around at the burned-flat apartment building.

"This stuff," he said to Ryan and Dave, meaning home-cooked meth, "is just kicking the hell out of us down here."

CHAPTER TEN

In the Black Museum

William Petersen as Gil Grissom

The microscopic debris that covers our clothing and bodies are the mute witnesses, sure and faithful, of all our movements and all our encounters.—Edmond Locard

Blood is jolly.—Alfred Hitchcock

The death struggle was confined to the kitchen. Eric Grimes, an elderly African-American male afflicted with high blood pressure and diabetes, lived next door to his younger brother, George, in Calumet City, near the Indiana state line. George dropped by Eric's small, orange-brick Foursquare, identical to his own and to six others on the block, at midnight on a winter Tuesday. Calling out his brother's name, he entered the kitchen and saw blood, great splattered gouts of it, on the floor and ceiling and across two of the walls. Eric's dead body lay on the floor next to a heating vent.

Upset and fearful that the attacker might still be in the house, George left the scene immediately and went back next door to call the police. By noon on Wednesday, homicide detectives had released the scene, and early the following morning, I pulled up in front of Eric Grimes's house with Dave Creager and Ryan O'Shea. The bitter cold made changing in the back of the truck a hurried process. Cloud-colored frost coated the curb and the boulevard strip.

George Grimes, a heavyset fifty-five-year-old former athlete whose movements were slowed by obesity and arthritis, emerged from his own house with the key to his brother's.

"I ain't been inside, not since I found him," he said to us. "My constitution isn't strong."

But he unlocked the main door, on the south side of the house, sheltered by a small portico. He hung back as Dave and Ryan did their initial walk-through. Hesitating, he moved down the brown-carpeted hallway toward the dining room. Through a doorway to his left was the kitchen. I watched as George inched forward and finally allowed himself a peek. Shock, dismay, anger, and sadness passed across his face in successive waves. He shook his head.

"Lord, he must've struggled," he said, staring at the curtain-like sweeps of blood reaching high above the refrigerator on the opposite wall.

The body had already been removed, but the fluid imprint of it remained.

Ryan steered him away from the kitchen, back down the hallway toward the living room. "What we're going to do, Mr. Grimes," he said, "is deal with all the blood and biomatter that's in the kitchen, remove whatever is contaminated. Then we biowash the walls, which is a three-step process."

Ryan looked down at his clipboard. "We're going to have to take up part of the carpet in the living room, too, but we'll save as much of that as we can."

Dave came up from basement and motioned to Ryan. The two of them huddled briefly. Ryan turned back to speak to George.

"Okay, it looks like there's quite a bit of contamination downstairs," Ryan said. "We're going to have to deal with that too."

That was it for George Grimes. "You do what you got to do," he said. Tears welling in his eyes, he backpedaled down the hall to the front door.

What Ryan had termed *contamination* in the basement turned out to be extensive. Fresh blood had drained from Grimes's head wounds, into the heating vent next to where he lay, down through the ductwork, finally to drop onto the basement floor. Over the course of the few hours before George discovered Eric's body, the steady drip of blood produced a perfect halo of splatter that was six feet in diameter and covered a chest freezer and a wheelchair stored in the basement.

"He rolls six inches the other way, away from the vent," Ryan said, "this job would be half as big."

Blood makes up less than ten percent of body volume. In the pool in the kitchen, I again noticed the separation that occurs when blood ages in open air. Fifty-five percent of blood is plasma, a saline solution that contains every protein produced by the body, more than five hundred of which have been identified so far. Plasma also contains several minerals and sugars, and is responsible for the "sweet" smell of blood so often mentioned by thriller writers. The pool that bled out from Eric Grimes had halfway dried and coagulated, and pink-tinged plasma had separated out on its perimeter.

I noticed something else about the scene. Even though it had been processed by homicide detectives, there was little residual fingerprint powder or other signs of the crime scene investigation. The scene was released with what appeared to be unseemly haste, a mere twelve hours after the discovery of the body.

"Think about it," Dave said as he geared up. "It's the South Side."

The issue of race hung in the air, unspoken. I imagined that there could be a lot of reasons why the scene was processed quickly, or perfunctorily. Perhaps the detectives had already developed a theory of the crime, or held a suspect in custody who had confessed.

Dave ripped open a pack of white terry-cloth rags and set to work on the hula hoop–size bloodstain on the kitchen floor. Ryan and Dave had decided the contaminated ductwork would have to be taken out and disposed of. While Dave worked the kitchen, Ryan and I went downstairs. The blood spatter in the drip zone below the ductwork looked so fine it could have been made with a plant mister. A half-moon of red droplets spread over the pitted white metal of the freezer.

I edged around the mess toward the corner of the basement opposite the stairs. A door led to a sheltered space outside beneath a deck. The door had a rusted security gate, a padlock hanging open on its catch. The security gate opened inward, and had scratched a powdery white arc in the gray concrete floor of the basement.

"Look at this," I said to Ryan. "I thought people had to lock their doors around this neighborhood."

"If it was a break-in, whoever it was could have gotten in right through there," Ryan said.

At nine o'clock on a dead winter night, the trespasser squeezed through the unlocked security gate, crossed the darkened basement, and surprised Eric Grimes in the kitchen at the top of the stairs. The murder weapon? One of opportunity, a hammer picked up from the clutter of the living room, perhaps, a metal cane, or the base of a lamp. Taken along when the murderer fled.

Ryan, who I've seen manhandle king-size mattresses into the back of the Aftermath truck all by himself, lifted the freezer up, rested it on his knee, and bent his head to look under it. He let it drop back to the floor with a bang.

"What the fuck is going on?" Dave shouted from upstairs in the kitchen.

"Nothing," Ryan shouted back. "Hollywood just fainted."

"I'm okay," I yelled up.

Ryan retrieved a roll of duct tape from his tool bucket and taped the lid of the freezer shut. I could hear the contents thudding around inside. Without asking for my help he upended the two-hundred-pound appliance and dragged it away from the blood halo.

"I think we can get away with cleaning this thing off and saving it," he said, and tossed me a package of cleanup rags.

....

Given a known set of circumstances—a cellar door left open, a blood-splattered kitchen—how do we tease out the details of what happened? Standing at point B—the crime scene—how do we journey back through time and arrive at point A—the crime?

In my journalism career I had repeatedly encountered the bafflingly opaque nature of the crime scene. I'd had enough experience to realize I wasn't that good at plumbing its mysteries, not in the intuitive sense of a Columbo, say, or a Gil Grissom. I stood at point B and peered backward, but point A remained stubbornly lost in the fog.

"You want to call it?" says Catherine Willows to Grissom, inviting him to present his theory of the crime. But I never seemed to be able to call it. My deductive reasoning skills never measured up to the task. Variables and unknowns overwhelmed me. I was, without apology, a Monday-morning detective. I liked to read affidavits and court transcripts. That's how I figured out what had gone down.

One of the first true-crime articles I ever worked on concerned a mass murder that happened in the countryside just outside my hometown in central Wisconsin. At 5:00 A.M. on the
morning of July 5, 1987, a fifty-five-year-old named Kenny Kunz
came home after sleeping off a drunk in his car to find his family
massacred. His uncle, two aunts, and brother lay sprawled
around their rat's nest of a farmhouse, all of them shot, ballistics
would determine, with a .22 rifle. Kenny's mother, seventy-year-
old Helen Kunz, was nowhere to be found.

Four dead, one missing. Kenny was the immediate suspect
(look closest first). His alibi wobbled. During the night of July 4,
when his family was murdered—the pop-pop-pop of the .22 lost
amid the exploding firecrackers and cherry bombs of Independence Day—he had been asleep in his white Ford Grenada at his
place of work. And, yes, because this was Wisconsin, his place of
work was a cheese factory.

As the details of the investigation leaked out, the victims
proved themselves to be as strange as the crime. Family patriarch
Clarence Kunz lived with a small harem of sisters—Helen, Irene,
and the mildly retarded Marie. Hints of incest hovered around
the intensely secretive family: Kenny was Helen's son by her
brother, making Clarence both his uncle and his father. His
brother, Randy, routinely slept in the same bed as his mother,
Helen.

Investigators discovered over twenty thousand dollars in cash
hidden in the incredible squalor of the Kunz household. There
was a lot of porn too. The farmhouse didn't have running water,
but it was equipped with a VCR and multiple TVs. As sheriff's
deputies sifted through the scene the day after the killings, a UPS
truck drove up to deliver a porn tape addressed to the recently
deceased Randy.

Detectives could not quite take Kenny Kunz seriously as a viable suspect. He was too slow, too muddled. But he did point them to a young neighborhood hellion who was eventually charged with the crime. Christopher Jacobs III—his family called him Christy—had visited the Kunz farmhouse a couple years before, looking to buy some autos junked on the land.

Through painstaking detective work, police developed a case against Christy Jacobs. It was circumstantial, and built on what forensic experts term "impression evidence," meaning imprints left by one object coming into contact with another. In this case, .22-caliber cartridge casings recovered in Christy's bedroom matched those found at the Kunz farmhouse. Also, a quarter mile away from the farmhouse, the freshly tilled Kunz family garden plot showed tire tracks that matched those on Christy's black 1974 Dodge Charger. Christy's motive, detectives theorized, was robbery.

Almost nine months after the crime, detectives found Helen Kunz's decayed body in a swamp along a rural road. She, too, had been shot twice in the head with the same .22 rifle. The road where she was found was known to investigators as one along which car-fanatic Christy Jacobs sometimes discarded unwanted auto parts.

Four months later, Christy Jacobs was charged with the five murders of the Kunz family. Thirteen months after that, in October 1989, a jury acquitted him. Circumstantial impression evidence, in those pre-*CSI* days, wasn't enough to sway the jury.

I followed the case from the beginning and as it threaded its way through the justice system. My father sent me clippings from the local newspaper. Like a lot of people around my hometown, I put on my deerstalker cap and tried to figure out what had happened. But I remained stuck at point B.

The facts about the crime scene were well-established. Four dead in a ramshackle farmhouse. Another family member killed by the same weapon, but discovered almost twenty miles away. Telltale cartridge casings. And a quarter mile down the road, in the soft dirt of a freshly Roto-tilled garden, tire imprints.

What had happened? I tried, but I couldn't fathom it. The past remained impenetrable, opaque.

I had a lot of time to think about it. It would not be until almost twelve years after the crime, in June 1998, that the facts about what went down on the secluded Kunz farm that Independence Day night would finally become clear.

You want to call it?

....

The murder of Eric Grimes in Calumet City got me closer to a fresh unsolved homicide scene than I had ever been before. Ryan, Dave, and I entered the house hard on the heels of the crime scene investigators. If it had been Las Vegas, the odor of Gil Grissom's cologne would still have been lingering in the air.

Most people learn what they know about crime scenes from watching television, and thus much of what they know is mistaken, only half true, or downright wrong. The difference between the reel world and the real world is marked.

But it isn't for lack of information. Counting network and cable, at least one hour of crime-forensics programming airs in prime time on six of the seven nights in the week. Three prime-time iterations of producer Jerry Bruckheimer's *CSI* franchise are among the top-rated television shows in the U.S., and *CSI* is also seen on several channels in Europe, South America, Asia, and Australia.

CSI is just the tip of a very large iceberg.* "Crime," one television critic dryly noted, "is the new black."

Hollywood created the mythical character of the crime scene investigator, hybridizing the roles of homicide detective, evidence analyst, and forensic scientist into one sexy beast. Gil Grissom, Catherine Willows, and their counterparts in Miami and New York, Horatio Caine and Mac Taylor, act as real jacks-and-jills-of-all-trades. They collect evidence at the scene, interrogate witnesses, badger suspects, head back to the lab to crunch chromatographs. They solve crimes.

Members of a real crime scene investigation team—in the UK they are called "scenes of crime" officers—have very distinct roles. The white smock behind the microscope never conducts witness interviews. The crime scene analyst passes off his material to the lab and might never see it again, under a microscope or otherwise. The evidence tech operates apart from the microscopist. Several different detectives might work different aspects of the same case. The all-in-one model makes for good drama, but it bears little resemblance to what actually happens in the field.

Forensics. From the Latin for "in the forum," meaning in public discourse. Due to its wide exposure in TV and movies, the term has become synonymous with crime investigation, but that's not really accurate. The word can just as well apply to the practices of formal debating teams, the traditional geek refuge in

---

* *Forensic Files* represents a documentary approach to the same type of material, as does such A & E and Court TV shows as *Cold Case Files*, *The First 48*, and *Trace Evidence*. Other police procedurals and forensic dramas figure prominently in the prime-time network lineup: *Law & Order* (with two spin-offs, *Special Victims Unit* and *Criminal Intent*), *Cold Case*, *Bones*, *Crossing Jordan*, *Criminal Minds*, *Numb3rs*, *Navy NCIS*, and *Without a Trace*. Alongside popular American imports such as *NYPD Blue* and *CSI: Las Vegas*, homegrown British crime show productions have always been popular in the UK, from the groundbreaking *The Sweeney* through *Taggart* and up to *Prime Suspect* and *Life on Mars*. In Canada, *Da Vinci's Inquest* features Nicholas Campbell as a coroner in the lead role.

secondary schools. Forensic science, then, refers to science designed for debate in the public forum—in other words, in court.

But because of *CSI*'s popularity, forensic science courses, not debate teams, proliferated in high school and college curricula. The trend spills beyond education. A bar opened up on the Bowery in New York City that calls itself Crime Scene. It has come to this: In a series aimed at middle school readers called Extreme Careers, there was a book called *Forensic Scientists: Life Investigating Sudden Death*. Some of the glamour filtered all the way down to bioremediation. Also available in the same series: *Biohazard Technicians: Life on a Trauma Scene Cleanup Crew*. Light reading for young teenagers.

Forensic science is indeed an engrossing discipline, maybe not as compelling as the drama-inflated narratives of crime writers, but one with its own attractions. Taken to an extreme, the world-vision of forensic science can represent a Zen approach to reality. It's a world where our every move radiates consequence, where our physical actions can be exposed and tracked as if by an all-seeing God.

Did the killer of Eric Grimes enter through the unlocked downstairs door, walk through his basement, and attack him in his kitchen? Theoretically, forensic science could trace the killer's every step. Do we consider our secrets safe when we operate alone and unseen? Forensic science sees. Forensic science presents our passage through the world as one that leaves behind a boatlike wake of evidence, swirling microscopic clues, a phosphorescent path that could conceivably lead back to our first steps as a toddler.

That's theory. In reality, there are severe limits on the omniscience of forensics. Our collection and retrieval devices are too inefficient, and our analytical engines too ham-handed, to be able to determine more than a small percentage of the clues that are theoretically available. Processing remains arduous and time-consuming. And that old bugaboo, human error, constantly threatens to corrupt the purity of the results.

Still, the idea lingers. There are no secrets, just unexplored routes of discovery. Theoretically, every step that O. J. Simpson took on the night of June 12, 1994, left behind microscopic traces. A team of forensic scientists with infinite computing power, finely calibrated machines sensitive to the atomic level, and an endless amount of time could proceed from point B back into the past and arrive at point A.

The source of this omniscient view of forensic science resides in the work of a pioneer French criminologist named Edmond Locard. A thick-bodied graybeard who in 1910 founded the world's first crime laboratory in Lyons, France, Locard is much-beloved by the world's crime writers. His name has become a shibboleth and a plot device, often introduced in the same way.

"Do you know who Edmond Locard was?" She shook her head.
                                    —Jeffery Deaver, *The Bone Collector*

"Have either of you ever heard of a man named Edmond Locard?" I said no.
                                    —Stephen White, *Cold Case*

"I don't know how much you know about forensics," he murmured. "Not much," she admitted. "The French investigator Edmond Locard is often credited with being the father of forensics," Tidwell told her.
                                    —Laura Childs, *Chamomile Mourning*

Locard's name endures because a guiding precept of forensic science has been ascribed to him. Now confidently cited as "Locard's theory," "Locard's law," or "Locard's exchange principle," its actual wording turns mushy upon close examination. Locard's rule of thumb has been variously stated as "Every contact leaves a trace," or "With contact between two items, there will be an exchange," or "The criminal always leaves behind something at the scene of the crime and carries away something from the scene that was not on him or her before."

The confusion arose from the fact that Locard never exactly formulated the idea as a "principle." The closest he came was with a sentence in his 1923 treatise, *Manuel de Technique Policière:* *"Il est impossible au malfaiteur d'agir avec l'intensité que suppose l'action criminelle sans laisser des traces de son passage,"* which can be translated, "It is impossible for a criminal to act, given the intensity of a crime, without leaving traces of his presence."

From such slim basis (there are other, similar declarations in Locard's writings), the concept of Locard's exchange principle took on a life of its own, until an obscure nineteenth-century French criminologist found himself routinely referred to in procedural thrillers and on prime-time television.

How did that happen? Human beings want to believe in our impact upon the world, we want to believe that we leave an imprint, a trace, the indelible evidence of our existence. Locard's exchange principle enshrines this wish in science. It is not just the criminal, acting in the "intensity" of the crime, who leaves a snail trail of evidence behind. We all do. Comforting to think so, anyway.

The historical Locard trained in medicine and law and became a police prefect in Lyons. He proved himself an obsessive, dogged forensic scientist long before the field was formally recognized. He embarked upon a decade-long microscopic study of various kinds of dust. He and Gil Grissom would have gotten along well. After solving a string of spectacular cases, Locard became known as the French Sherlock Holmes, and his forensics laboratory the goal of pilgrimages by police all over the world.

Within his lab, on the fourth floor of the Palais de Justice in Lyons, Locard created what he called the Black Museum, a showcase of sorts, featuring murder weapons, evidence from famous homicides, and a photographic rogue's gallery of killers. The exhibits of the Black Museum represented a precursor, the very first pre-TV, pre–Jerry Bruckheimer protoepisodes of *CSI*.

The Black Museum had a distinguished visitor in 1921, when British author Arthur Conan Doyle stopped in Lyons on his way back to England from a sojourn in Australia. A strange funhouse-mirror moment must have occurred, when art met science, and the creator of Sherlock Holmes encountered the reputed real-life French personification of his creation. Locard conducted Doyle on a tour of the Black Museum.

Doyle stopped short in front of a portrait in the rogue's gallery. "Why, what is my chauffeur doing here?" Doyle asked.

"You're mistaken, Sir Arthur—that's Jules Bonnot, the famous motor-bandit."

But the villain in the rogue's gallery and the chauffeur were one and the same. Arthur Conan Doyle's driver in the years before World War I later became infamous as the murderer, carjacker, and "illegalist" Jules "le Bourgeois" Bonnot, credited with the first use of the getaway car in crime. Bonnot had worked not only for Doyle, but for some of the top forensic scientists of the day, a criminal-in-the-making gleaning pointers from leading experts in the field. He then fell in with a group of French anarchists to create *la bande à Bonnot,* the Bonnot gang, and was killed in a spectacular shootout with the gendarmes.

Taking the Black Museum as a starting point, and running up to the present-day saturation of prime-time television, we've had about a hundred years of forensics-as-theater. The cumulative effect of this onslaught has been to skew the public mind in some odd ways.

....

Prosecutors and defense lawyers label it the "*CSI* Effect," and it has already become a recognized element in American jurisprudence. Schooled by a steady curriculum of crime shows, the public formed new assumptions about the validity of scientific evidence. The jury pool became muddied, or at least clouded.

Predictably, prosecutors and defense attorneys disagree on the impact of the *CSI* Effect. Prosecutors detect highly unrealistic expectations on the part of juries, who anticipate DNA tests, trace-evidence analysis, and all the whiz-bang scientific bells-and-whistles common on television but not always applicable to every case. Prosecutors have begun presenting "negative evidence witnesses," summoning experts to testify that just because DNA results were not presented in court does not mean the prosecution's case is weak. Such negative testimony in court attempts to re-educate juries in the realities of criminal investigation, counteracting the perceived harmful impact of the *CSI* Effect.

Likewise, defense attorneys have sought to neutralize an assumption they ascribe to too much TV crime viewing on the part of juries. On *CSI,* the evidence never lies and Gil Grissom is never wrong. As a result, juries have developed an unrealistic faith in scientific evidence, which they perceive as infallible, just like on TV.

"You never see a case [on TV] where the sample is degraded or the lab work is faulty or the test results don't solve the crime,"

Dan Krane, president and DNA specialist at Forensic Bioinformatics in Fairborn, Ohio, told *USA Today*. "These things happen all the time in the real world."

The *CSI* Effect represents a subtle variation of the Heisenberg uncertainty principle, which maintains that the act of observing changes that which is observed. The act of portraying crime—on television, in movies, in the pages of this book—changes its nature. To put it another way, art fucks with life, and life turns around and repays the favor by fucking with art.

A young teenager fishing off his family dock in Galveston Bay saw an object floating in the water and called to his father, who judged that it was the body of a dead pig. The floater turned out to be a headless and limbless human torso, and its discovery eventually led to murder charges against New York real estate heir Robert Durst. The same day, September 30, 2001, police examined garbage bags washed up with the tide nearby. The bags contained the torso's arms and legs (but no head) as well as convenient clues to the identity of the dead man. He was Morris Black, Durst's seventy-one-year-old neighbor and the erstwhile owner of what would become the most infamous missing head this side of Sleepy Hollow.

As the scion of a Manhattan real estate fortune worth billions, Durst could afford the best legal representation money could buy. (A pot-smoking, cross-dressing eccentric, Durst petitioned Judge Susan Criss to fire his attorneys, writing that he "had already paid $1.2 million in retainers" to his legal team, but it was now trying to "squeeze him for another $600,000.") Among the things that $1.2M can buy are the services of famed jury selection consultant Robert Hirschhorn, taken by some to be the model for the Gene Hackman character in John Grisham's *The Runaway Jury*.

Hirschhorn was very well aware of the *CSI* Effect and its implications for his clients. The Durst defense wanted to play up the fact that a missing piece of vital physical evidence—namely, Morris Black's gone-astray head—might help exonerate the accused. On shows such as *CSI*, Hirschhorn knew, such a major gap in evidence was unheard of, and might lead jurors to judge the prosecution's case weak. Durst's legal team insisted the case was one of simple self-defense. Wounds to Morris Black's head, they said, would prove out the theory, if only it could be found.

Hirschhorn made sure prospective jurors were surveyed as to their TV-viewing habits. He didn't have to worry. Out of the five hundred citizens in the Galveston jury pool for the Durst trial, more than two thirds watched forensic crime programs such as *CSI* or *Law & Order*. In a finding that ranks with the "what were they thinking?" outcome of the O. J. Simpson trial, the jury acquitted Durst of killing Morris Black even though he had admitted doing so, and had confessed to chopping up the body with a butcher's saw too.

But it's not only prospective jury members who watch TV. Killers do too. Police and prosecutors have noticed a trend recently toward generally less physical evidence recovered from crime scenes. Sitting at the feet of Gil Grissom and Andy Sipowicz, felons have learned to cover their tracks better, scour for trace evidence more completely, and defeat Locard's exchange principle as best they can.

Perpetrators spill a lot more bleach around crime scenes nowadays. Torching evidence is also more common. Rapists, attuned to the incriminating potential of DNA, have begun to use condoms much more often than previously. Thus was born a new category of trace evidence, whereby scientists analyzed and matched particulates, lubricants, or spermicides to particular brands of condoms. In a parody of an arms race, the light and dark sides of the Force competed to develop new ways to apply Locard's law, and new ways to circumvent it.

On a Wednesday evening just before Christmas 2005, in Warren, a suburb of Youngstown, Ohio, Keyatta Hines, twenty, acted as a driver for a dark errand. With her were a forty-five-year-old sometime dope dealer named Rebecca Cliburn and a young ex-con named Jermaine McKinney. Already in his young life McKinney had been given the nickname "Maniac," and he was going to earn it that night.

The plan: Hines would drop McKinney and Cliburn off at the house of Cliburn's mother, Wanda Rollyson. Rollyson was off at church and the house was, McKinney would tell Cliburn, a perfect place for them to have sex. Hines would then speed off to pick up another partner in crime, Jazzmine "Jazz" McIver, twenty-one. The two girls would then come back to the Rollyson house to help McKinney rob Cliburn and her mother. As an added bonus, Wanda Rollyson was a diabetic, and always kept syringes around. In the world of Maniac McKinney, syringes were gold.

At 6:30 that evening, Hines dropped Maniac and Cliburn off at Rollyson's home on the outskirts of Warren, close to the Ohio Turnpike. By the time Jazz and Hines returned at around 9:00 P.M., events had already transpired well beyond robbery. Hines later told police she entered the house to find the elderly Rollyson back from church and lying dead in "big puddle" of blood. Cliburn was dead too. McKinney had fatally beaten her with a crowbar after putting two bullets into Rollyson's head.

Among his other leisure time occupations (marijuana, concealed weaponry), Jermaine McKinney was a devoted viewer of *CSI*. "I don't want evidence found," he told Hines repeatedly. He was concerned about hair, he said, and the sperm he had left in

Cliburn's body. He carefully collected his cigarette butts from the scene, knowing from *CSI* that they could yield DNA in his saliva. He washed his hands in bleach, confident that would eliminate traces of the crime. Covering his car seats with blankets to prevent blood droplets from being transferred, he gathered his bloody boots, the bed linens, and the murder weapon and brought them out to the car. Then he returned to the house.

In order to eliminate any trace evidence he might have left behind, Maniac decided Cliburn and her mother would have to be burned. He carried the battered bodies to the basement, painted them with flammable beige house-paint, and set them afire. In this step, at least, he was successful, since police later said they could identify the bodies only by the jewelry they wore.

His attempts to dispose of the crowbar and bloody clothing were not so successful. Driving through the night with Hines and Jazz back to his old 'hood in Youngstown, McKinney tossed the crowbar and other incriminating evidence off Jacobs Road into McKelvey Lake. Or rather, *onto* McKelvey Lake, since the weapon failed to break the ice and came to rest in full view, alongside his boots and Wanda Rollyson's blood-soaked bed linens.

"Motherfucker!" McKinney screamed, standing there freezing on Jacobs Road, looking down at his best-laid plans gone awry. A maniac, maybe, but not a smart one. Two days later, after police captured Maniac (true to form, he engaged them in a four-hour gun battle), investigators retrieved the murder weapon and the other discarded material.

"People are getting more sophisticated," an investigator with the Trumbull County prosecutor's office told the Youngstown *Vindicator,* "with making sure they're not leaving trace evidence at crime scenes."

A contrary opinion, from Larry Pozner, former president of

the National Association of Criminal Defense Lawyers: "Most people who commit crimes are not very bright and don't take many precautions," he said to the Associated Press, serving up commentary upon, but probably not directly referring to, Jermaine McKinney as he heard the crowbar clang down on lake ice. "*CSI* and all the other crime shows will make no difference."

....

Christy Jacobs told the story of the Kunz killings to a girlfriend of his who, when she was nabbed on a bank robbery beef in Minnesota, told the story to police. After a little detour to the Supreme Court to sort out double-jeopardy issues, Jacobs was tried for the kidnapping of Helen Kunz and the full story came out in court.

He went to the Kunz farmhouse that night not for burglary but for murder. After an argument with his father he felt angry and demeaned. The Kunz family seemed the perfect targets for his ire—a group of outsiders living in an isolated location. An opaque element of the crime scene, at least to me, was always the car parked a distance away in the garden plot. That turned out to have a very simple explanation, very obvious in hindsight: Jacobs wanted to sneak up on the house. The broken arm that Randy Kunz suffered before he was shot stemmed from a struggle with Jacobs at the door of the farmhouse.

Still sexually inexperienced at age twenty-one, Christy Jacobs kidnapped the seventy-year-old Helen Kunz in order to rape her and thus finally unload the burden of his virginity. He drove with the terrified woman through the dark Wisconsin countryside. When it came time to do the deed, Jacobs pulled over to the side of the road. But he couldn't do it. He opened the car door of his

Charger and vomited onto the asphalt. Then he took his passenger from the car, sat her up against a tree and, using the same Remington Nylon 66 .22 rifle he had employed to kill her brother, son, and sisters, shot Helen Kunz twice in the head.

Yes, of course. That's exactly what must have happened. For over a decade I had stared at the opaque screen hiding the facts about the Kunz farm massacre. Then the veil lifted, and a sense of clarity flooded the crime with light. Why hadn't I been able to parse it out before? But that was the nature of the beast, and the attraction of forensic science. Humans are the puzzle-solving animal, and crime scene investigators solve high-stakes puzzle after high-stakes puzzle professionally, in the course of their ordinary workdays.

But my crime-parsing skills were progressing. Two weeks after the murder in Calumet City, detectives arrested an ex-con junkie named Chessie Martin, who confessed to killing Eric Grimes. He had targeted the same neighborhood in a series of break-ins, always accessing unoccupied houses through their back alleyway entrances. Grimes surprised Chessie Martin in the darkened kitchen before he ever got a chance to rifle through the house for valuables.

"He fought hard," Martin told detectives. "I thought he was going to kill me, so I had to kill him."

CHAPTER ELEVEN

The Left Hand
of God

The Freak show in Plainfield, Wisconsin

Reporters love a murder. —Calvin Trilling

I'm a killer-diller with nothing on my mind.

—Fats Waller

Notes toward a murder biography, or, Homicides I Have Known and Loved.

Ed Gein enjoyed considerable notoriety around the central Wisconsin of my youth as a hometown boy who had made bad. He lived in a ramshackle farmhouse outside of tiny Plainfield, Wisconsin. (Later on, it wasn't lost on me that Aftermath's headquarters was located in Plainfield, Illinois.) Locals employed Gein as a handyman and occasional baby-sitter.

Baby-sitter? Evidently so. There may be people still alive who as children were dandled on Ed Gein's knee. Even before he twisted off, Ed was odd. What could their parents have been thinking?

When his mother, Augusta, died in 1940, the thirty-nine-year-old Gein became gradually but thoroughly unhinged. He had always been something of an authority on anthropological studies that showed odd or bizarre human behavior. Now he began to collect anatomy books, medical encyclopedias, reference works on the Nazi experiments at Auschwitz. His gender dysphoria led him into strange confusions, and he investigated sex-reassignment surgery but then held back. He decided instead on a do-it-yourself approach. With his semiretarded grave-digger sidekick, Gus (last name lost to history), he made body-robbing sojourns to local cemeteries.

Gein's prey were always adult females. He dug up Augusta herself. Using the meat-dressing skills he had learned during deer-hunting season to flay and cure human skin, Gein fashioned the corpses into macabre trophies. When police finally showed up at the farmhouse in November 1957, they found a shoe box full of nine salted vulvas (one, presumably that of his mother, painted silver), a belt embedded with nipples, four noses in a cup in the kitchen, and a pair of lampshades made of human skin.

The discovery happened a month after my fourth birthday. With only two confirmed kills, Ed did not technically qualify as a serial killer (the traditional minimum requirement was three), but that didn't deny him immediate entry into the pantheon of folk mythology. Ed Gein jokes were a constant of my childhood (Q: What does Ed Gein eat for dessert? A: Ladyfingers). Wisconsinites became known as "Geiners," which was, I suppose, a better tag than "cheeseheads."

In the aftermath, storytellers began to recast Gein into some of the most infamous horror characters of the movies. Norman Bates in *Psycho,* Jame "Buffalo Bill" Gumb in *The Silence of the Lambs,* Leatherface in *The Texas Chain Saw Massacre*—all were incarnations of Ed Gein. He became that flip side of the folk hero, the folk villain.

Plainfield was forty miles south of my hometown. My father often drove there on business. A carny named Bunny Gibbons bought "Ed Gein's ghoul car" and toured area carnivals with the 1949 Ford as a twenty-five-cent novelty draw ("Look! See the car that hauled the dead from their graves!"). I paid my two bits with my friends at the Wisconsin Valley Fair.

Gein's gravestone in the Plainfield Cemetery led a peripatetic existence. After years of being chipped away at by souvenir hunters, it finally disappeared entirely in 2000. A year later it turned up across the country, outside Seattle. It now rests in a museum in Wisconsin. The Gein farm had a similar twisted history. The farmhouse itself burned to the ground on March 20, 1958, the work, police said, of arsonists.* In 2006, the Gein land briefly came up for sale on eBay. ("Ed Gein's Farm . . . The REAL deal!") The ad got a lowball bid before being pulled by the auction site's administrators, who had banned the sale of so-called murderabilia in 2001.

The jokes, the movies, the continued fascination were no doubt expressions of communal nervousness. How could anyone stroll off the deep end in quite that weird a way? Salted vulvas?

I eventually came to see Ed as the inevitable product of the normalcy of my home state. What could be more banal, mundane, and average than the Midwest? But there's a price to be paid for all that dullness. Every so often an excrescence of the macabre had to come along to balance things out. Ed Gein and then, thirty years later, Jeffrey Dahmer. The state should be due for another monster along about 2020.

Because of its shape on the map, Gein called Wisconsin the "left hand of God."

Ed's weirdness placed him beyond the pale. He was risible rather than hair-raising, and the proper response when his name was invoked was a nervous chuckle. Layered underneath the laughter and anxiety, I thought I could detect a sympathy, almost an affection.

---

* Ed Gein's farmhouse, destroyed by fire. John Wayne Gacy's house in Des Plaines, bulldozed flat, as was O. J. Simpson's Rockingham estate. Jeffrey Dahmer's Oxford Apartments building, with its infamous Apartment 213, demolished. These sites were, in the language of real estate experts, "stigmatized" properties. You know you have transgressed in some basic, elemental manner when authorities raze your house and sow your fields with salt.

Some of this was just gallows humor. Whistling past the grave-yard. Students at my alma mater, the University of Colorado, named the cafeteria in the student union the "Alferd E. Packer Grill," in honor of Alferd Packer, a Colorado gold prospector who in 1883 became the country's only convicted cannibal.

Better to store a monster safely away in jokes and mythology than have him roaming around the streets. In folklore, Gein assumed the aura of the holy whack-job. His craziness had the soothing effect of making everyone else appear sane. He was a poet of carnival madness. Wearing a "mammary vest" stitched out of breasts and leggings made of human skin beneath his dead mother Augusta's dress, he danced outdoors at night.

"Graze the skin with my fingertips," he recalled of these Dionysian rites, "a pleasant fragrance in the light of the moon."

What can you say about a man like that? For all of Ed's noto-riety during my childhood, he might have been too bizarre to be truly frightening.

No, the first murder that came truly, violently home to me occurred at around noon on Monday, May 11, 1964, when I was ten years old. A former mental patient repeatedly stabbed an eighteen-year-old high school senior named Eleanor Kaatz as she crossed a railroad bridge over the Wisconsin River in my home-town. The island where it happened was one of my early play-grounds, officially called Barker-Stewart, which we always referred to as Dog Island. A deserted, ruderal-strewn place, bisected by railroad tracks, perfect for childhood play, or for murder.

I had crossed the "murder bridge" (as it instantly became known) many times. My childhood friend Brian and I once built a makeshift fort a few yards from where searchers discovered Eleanor Kaatz after the attack. Volunteers in a fire department rescue boat rounded the northern end of the island and saw her, sitting slumped on the riverbank, dazed and in shock, covered in mud. Her feet trailed in the water and her gingham dress was bloodied, with more than fifty jackknife stab wounds in her face, neck, back, and abdomen. The stubby blade had punctured her liver a half-dozen times. She bled surprisingly little externally, but had massive interior hemorrhaging. To the astonishment of her doctors, Kaatz held on for nearly two days before succumbing.

The world shifted. I was badly frightened. To my ten-year-old eyes the light off the river appeared somehow altered, to become lowering, starker, more menacing. When Truman Capote published *In Cold Blood* two years later, and when I got old enough to read it, I related precisely to the way he described the yellow flare of the flashlights and the blue muzzle-blaze of the shotgun, as they illuminated the deaths of the Clutter family. The light that shows murder shines very strangely, because it is filtered through fear.

On Tuesday evening, the day after the attack, sheriff's deputies cornered the former mental patient, twenty-one-year-old Terry Caspersen, in a woodlot that was, because this was Wisconsin, near a cheese factory. He was "dirty and bedraggled," police said, from sleeping the night in the woods. Caspersen, who had spent the year previous in and out of Winnebago State Hospital (aka the Northern Asylum for the Insane), had a rock in his hand when he confronted Deputy John Luebbe.

"What's that rock for?" Luebbe shouted, drawing his service pistol.

"I want you fellows to kill me," Caspersen answered. Then he dropped the rock and went meekly.

Caspersen had spent his childhood morbidly shy, so fearful of criticism that he failed third grade after an offhand remark from a teacher devastated him. He became a car thief and a firebug, inserting kerosene-soaked rags beneath the asphalt shingles of

houses and lighting them up. In Duluth, where he lived for a period, he burned down seventy buildings, most of them occupied
residences, but managed to kill no one.

While in custody, Caspersen heard the jail radio announce
the death of Eleanor Kaatz, and the blood drained from his face.

He gave police an account of the killing. He went to Dog Island, he said, in order to end his life. He felt depressed that Monday. Homicide, the shrinks tell us, is misdirected suicide, and
Caspersen's impulses morphed. He concealed himself in the
brush beside the bridge. He let an elderly woman trudge by.
Then came Kaatz, heading downtown for lunch with a friend.
He emerged from the woods and grabbed her.

"I know who you are," Kaatz said coolly, and in fact the two
lived nine blocks away from one another, though Caspersen told
police he had never seen Kaatz before. Caspersen herded her
ahead of him across the bridge and into the swampy brush of the
island. He then tripped her from behind and began to "pound
away at her with a knife" in the memorable words of a local newspaper account.

After the frenzied assault, Caspersen left the scene. He then
remembered that Kaatz had a transistor radio in her purse and
went back for it, but fearing a fingerprint ID tossed both the radio and the murder weapon into the river. He spent the rest of
the day making calculated appearances in various parts of town,
attempting to establish an alibi. But after a little more than
twenty-four hours on the lam ("running," he called it, although
he didn't run too far), he encountered Deputy Luebbe, and lost
their impromptu game of rock-paper-scissors-gun.

A murder occurs down the street from your house. Your
brain goes into overdrive, and a hundred images flash through

your mind—the blood, the knife, the scene of the crime. A couple of these images get stuck in your head forever.

Why? What makes murder memorable? The most obvious reason is the threat, the fright, the close call, however imaginary. The antelope always remembers the hot breath of the lion. But more broadly, murder is a very big deal. In *Star Wars* terms, it is a disturbance in the Force, in a way that defines the whole idea of a disturbance in the Force. Why does Hamlet hesitate? Isn't it obvious? Murder is huge. It isn't a simple decision. Premeditated killing is enough to make even a moody, broody Dane balk and think twice.

I don't remember the name of my fourth-grade teacher, a woman I spent almost every day with for nine months when I was ten years old. I don't remember my birthday party that year, or what presents I got for Christmas. But I remember my first murder, and the image of the girl in the bloody gingham dress sitting by the riverside.

. . . .

Perhaps it was the evil eye of Ed Gein, or the lingering influence of Terry Caspersen, but when I started writing, I started writing crime. It became, for me, the literary equivalent of worrying a bad tooth. I gravitated toward my fears, which were manifold.

Not until I wrote about the Susan Smith murders in Union, South Carolina, did I grapple directly with how my obsessions might affect others around me, particularly my wife and family. It would be difficult for anyone not around for those days in October 1994 to grasp how completely the Smith saga grabbed the country. Susan Smith murdered her children, allowing her Mazda Protegé to sink into John D. Long Lake with three-year-old

Michael and fourteen-month-old Alex strapped to their car seats inside. Then she lied about it, conjuring up a story of an African-American carjacker who stole her Mazda and her babies.

A year after the event, my wife and I helped Susan Smith's estranged husband, David Smith, Michael and Alex's father, write a book. We found David to be a straightforward young southerner who had been blindsided by an absolute shit-storm of grief and celebrity. I was convinced that if anything remotely similar happened to me, I would not be able to face it with the kind of forbearance that David Smith demonstrated. In fact, David's stoic dignity gave me a good feeling about the American character itself. If a South Carolina grocery clerk could hold up that well in a white-hot emotional crucible, then maybe not all was lost.

Doing research for David's book, we sifted through a storage locker's worth of letters, mementos, and gifts sent to him from all over the world. They were heartbreaking. Many of them used the tragedy of Michael and Alex to reopen old wounds of the writer's own. *Last July my own daughter drownded. . . . I made it through and I pray you do too.*

"Listen to this," David said to us one afternoon, while we were interviewing him in his apartment. He slotted a tape into his cassette deck. The quavering voice of an old blues singer came out of the speakers, singing a self-penned murder ballad about Susan, Michael, and Alex.

Oh, the mother, in the darkness
Oh, the mother, in the darkness
With her children by the lake as deep as hell.

"That guy sounds like a million years old," David said. His tastes ran more to Elton John.

Our daughter was still an infant when we began working with David Smith. We would fly down to Union, spend a few days interviewing him about the horror of infanticide, then fly back home and rush to check on our baby. I felt guilty. Maybe my flirtation with the dark side would somehow bring darkness into our own lives, into the life of my innocent daughter. Who knew what forces could be unleashed? A crime writer ought to be a lone wolf, responsible for no one's emotional demons but his own.

Or so I quite grandly thought. Where the experience of delving into Union's seamy underside prompted me to question my own morality, it served to fix the high moral compass of my wife. "That poor guy," she said repeatedly. "Those poor babies." Compassion, empathy. These are the only proper confessable responses to the pain and tragedy of others.

In his book of crime scene photographs, *Evidence*, Luc Sante expressed the compromised status of the crime writer, when he wrote of the photos: "I cannot mitigate the act of disrespect that is implicit in the act of looking at them." There didn't seem to be any way of getting around it. Examination of people's lives in extremis inherently violated the integrity of those lives.

I searched for solid ground in all this. Was I just feeding the milling hordes who waited outside the Union courthouse for a glimpse of Susan Smith in prison orange? Was the voyeur always culpable? If he was, then as a culture we were embarking upon a morally fraught course. Via television we had become a nation of voyeurs, and increasingly, the object of our fascination had drifted from sex to death. The more graphic, the better.

As recently as the nineteen eighties, an obscure cult "shockumentary" called *Faces of Death* marked the far edge of popular entertainment. The video series, serving up real and re-created footage of autopsies, beheadings, and violent death, maintained a furtive profile, marketing itself strictly among horror aficiona-dos. Now, of course, the spectacles displayed in *Faces of Death* have gone mainstream and prime-time, courtesy of *CSI* and other forensic-oriented programs.

Was the whole culture becoming more and more inured to death? The Moral Compass didn't think that was exactly the main point. "People used to have a lot more intimate experience with dead people," my wife, the history writer, told me. "They used to dress their own dead in their own households. So maybe we are just coming back to where we once were."

In parts of the world, people still dress their dead themselves. And in some other places the ancient intimacy with death and the new technology mesh in gruesome ways. On jihadi videos, a sta-ple of some Islamic communities, executions by the Taliban, be-headings and hangings, are the main draw. Bodiless faces of death.

In the course of working with David Smith, I tried to find my way out of the voyeur-as-perpetrator dilemma. I came to put my faith in the need most people feel to have their stories told. Many times it went beyond a need to an obsession, really. David Smith had solemnly promised Katie Couric on national televi-sion that he never wanted to make a penny off the death of his sons. Yet here he was, writing a book. His urge to sift the chaos that had engulfed him and extract meaning from it was just too strong.

I noticed the same urge again and again as I went on to write a series of simple four-thousand-word crime stories for *Maxim*. Hu-mans are the storytelling animal. During an interview with an in-carcerated parricide in an upstate New York county jail, I could not help but conclude that, rationally, this person should not be talking to me or any member of the press. It was not in his best interest to do so. Yet he blathered. He was compelled to do so by an inner urge to extract order from the chaos of his biography. I could feel him searching, as we talked, for some concatenation of words that would give his story shape.

"There is no agony," said the poet Maya Angelou, "like bearing an untold story inside of you." It's an anguish that trumps even deep grief.

Aftermath serves the same function, in a physical sense, that a crime writer serves in the literary sense. Chris and Tim are gatekeepers. They assist in the transfer between two worlds. One world is the crazed, anarchic, abnormal world of violent crime and tragedy. What Aftermath does is take that world and cleanse it of chaos, gradually transferring it back into tamer, more approachable reality.

A crime writer seeks to do the same thing. From the messy, random, loose-strings-untied world of violence, I always try to tease out a narrative. Order from disorder.

A Grateful Dead song called "Ripple" contained lyrics that Tim sometimes quoted in late-night bull sessions in the aftermath of Aftermath, when the physical element of the work was done. "Ripple in still water," went the simple refrain, "when there is no pebble tossed, nor wind to blow . . ."

That's what Chris and Tim think about their jobs, not the rock tossed into the water itself—since by the time they get there the rock has often completely disappeared into the murky depths—but the ripples left behind. Robert Hunter and Jerry Garcia's song might be about the hand of God passing through our lives. But human hands leave ripples as well. Chris and Tim encounter victims but never see them. They only experience the fact of their absence.

A good metaphor for this is the chalk outline that police sometimes draw (though not as often as TV crime dramas would suggest), silhouetting the position of a homicide victim. Indeed, getting "chalked" is street slang for getting murdered. Aftermath techs never get to meet the living, breathing human. They experience only the outline, the halo, the empty space that the victim's leaving had left in the world.

. . . .

From Ed Gein to Terry Caspersen to David Smith and *Maxim*. And finally my murder biography led me to Aftermath.

The job Greg and Greg took at the end of July that year was one nobody wanted. They arrived at the site, a duplex located just four blocks across the state line in Dyer, Indiana, on a blisteringly hot day. They met a relative who opened the house for them, and entered into a grisly, week-old double murder crime scene. Blood had spilled in both the dining area, off the kitchen, and the carpeted living room. But blood spatter trailed through the whole house, in the family room, kitchen, front hallway, and garage.

The blood spatter, by that time long dried and considerably blackened, served as a mute record of the crime. Greg and Greg could follow as the murderer chased a victim through the house, starting in the garage, through the hallway, into the kitchen. Most disturbingly, four-inch handprints on the dining area cabinet, and somewhat larger ones on the wall of the hallway, indicated the youth of the deceased. The victims were children.

The killer was their mother. Greg and Greg had only unwillingly taken the job because it had already received a lot of play in the press, and because Greg Sundberg had just become a new father.

"The strangest thing about Dyer was that you could play the whole killing in your mind just by following the trail of blood," Sundberg told me. "It was like you were there."

"Those fucking handprints," Banach said. "That's what got to me."

By most accounts thirty-year-old Magdalena "Maggie" Lopez doted on her two sons. Erik, a toddler, still wore diapers at age two, but his big brother, Antonio, who was also called Anthony, was about to celebrate his tenth birthday. His father, Robert, and he planned to watch Antonio's beloved Cubs at Wrigley Field. The two Lopez boys wore matching white cowboy hats as Antonio chased a shrieking Erik around the duplex's yard.

The Lopez marriage hit rough patches. Maggie had filed for divorce twice, Robert once, but they always reconciled. Police showed up at the house in 2001, summoned by Maggie during a high-decibel argument, but no action was taken after Robert agreed to spend the night away from home. Not totally smooth, but not entirely rocky either. Most of the time the Lopezes appeared to be a happy, well-adjusted family.

A half year after Erik's birth, in mid-2004, Maggie suffered a miscarriage that triggered a change in her personality. At times she became moody and mercurial. Other times, when she was home alone with the boys, she would retreat to a corner of the living room and sit there, immobile. Erik didn't understand. He wasn't old enough to be disturbed by his mother's strange behavior. Antonio, though, was badly frightened.

At the beginning of 2005 Robert's mother, Irene, took her daughter-in-law to a series of doctors for tests. Postpartum depression, one said. Another diagnosed bipolar disorder. They wrote prescriptions for psychoactive medications, Prozac and Xyprexa. The drugs seemed to smooth out Maggie's edges, calming her somewhat.

On the night of Tuesday, July 19, the very thing the drugs were designed to prevent occurred: Maggie experienced a break with reality. She would tell police later that she felt compelled to do what she did, that she could no longer take care of her children and that she needed to send them to a better place.

Grasping a ten-pound free-weight barbell, she attacked Antonio in the garage first. He ran from her, seeking refuge inside the house. She caught him once again in the front hallway, and finally administered a fatal blow as her son collapsed on the living room floor. Then she turned to the baby, howling in the kitchen.

At 10:59 that evening, Maggie placed a call to 911, telling the dispatcher she had harmed her sons. Officers arrived to find her walking barefoot out of the front door of her house, blood on her blouse and slacks.

"I wanted to send them to heaven," Maggie told detectives. "They needed to be saved. I was sick and couldn't take care of them anymore."

. . . .

Strictly speaking, and although they are commonly used interchangeably, murder and homicide have different meanings. The word *homicide* makes no judgment, and merely describes the fact of one person killing another. Such an event does not become "murder" until it is judged so in a court of law. A homicide can become many different things in court: manslaughter, justifiable killing (mostly self-defense), negligence. *Murder* is a legal term that invokes a strict set of conditions, such as premeditation and intent. The law requires *mens rea*, a guilty mind, which requirement allows all sorts of judicial mischief. It is why psychologists so often find themselves testifying in court.

There were 806,316 homicides in the U.S. between 1965 and 2004, averaging out to around 20,000 per year. About fifteen percent of reported homicides were deemed by the courts to be something other than murder. The U.S. murder rate—the number of murders per 100,000 people—has undergone a recent decline, attaining levels not seen since the early sixties: 5.5 in 2004, which represents 16,137 total murders. By contrast, England and Wales experienced 1,045 murders in 2002, a number that includes manslaughter and infanticide, for a rate of 1.41 per 100,000 of population.*

At the other end of the spectrum, neither the U.S. nor the UK rate comes anywhere near the predations of the Gebusi tribe of New Guinea. If we can believe the anthropologists, the Gebusi have the highest murder rate on the planet, an astonishing 568 per 100,000. Source of the trouble? Warring over biologically viable females.

Murder is overwhelmingly a male purlieu—men who murder outnumber women who do so by a factor of around nine to one, and murder is the tenth leading cause of death for males. Among young males in some sectors of the inner city populace, it is the leading cause of death.

But we are not alone. Murder is not solely a human endeavor. Researchers in Africa have witnessed chimpanzee murders, with cooperating gangs carrying out preplanned (and thus,

---

* Historical rates have been estimated as generally higher: 23 per 100,000 of population in thirteenth-century England, 45 per 100,000 in fifteenth-century Sweden, and 47 per 100,000 in fifteenth-century Amsterdam.

premeditated) killing rampages against individual rivals. The favored chimp method is skull crushing, with pulling off the fingers of victims as a preliminary tactic. Infanticide, too, is not unheard of, not only in the chimp world, but in many fiefs of the animal kingdom.

Forensic investigators, considering the variety of weapons involved in murders, have established broad rules of thumb that can help illuminate the nature of a killing. Blunt-weapon use, such as in the Lopez murders and the killing of Eric Grimes, can indicate a crime of impulse or passion. Caught up in the frenzy of the moment, the perpetrator seizes the nearest object. Blunt-object homicides are often, but not always, unpremeditated. Contrast the Lopez case, for example, with the Tarell Koss murders. Koss had to wait out a three-day cooling-off period before he could obtain his gun. In general, use of firearms indicates a greater probability of premeditation.

I never went out to the Lopez duplex. Instead I viewed Greg and Greg's digital-camera documentation of the scene on the computer monitor in the Aftermath offices. Aftermath, Inc., maintained an enormous database of crime scene photos, clinical representations that served primarily documentary (rather than narrative) purposes.

The techs marked the files with short descriptive phrases: "Suicide in basement," "Murder-suicide 30s couple," "Hemorrhage," "Murder stalking." Other tags referenced geography: "Elmhurst," "W. Walnut," "Roger Road," "S. Oak Park." A few of the descriptions rise to the level of minioperas: "Suicide cross-dressed male, wife in Hawaii," and "Unattended death, 9 days, male 50s, reading *Playboy* on crapper."

A pair of storage boxes containing Aftermath's pre-digital-era photo collection, thousands of depictions of jobs from the early days of the company, disappeared from its offices, the work, no doubt, of souvenir hunters. Crime scene photos have become popular on the Web, with no Luc Sante–style disclaimer about the "disrespect" inherent in looking at them. They have become just one more subgenre in a burgeoning media culture of crime. Because of the squeamishness of advertisers, *Maxim* had surprisingly stringent rules about crime scene photos, one of which outlawed blood flowing from body wounds. We could show blood, and we could show dead bodies, but the twain could never meet.

I sat in front of the computer in the Aftermath offices and scrolled through the images of the Lopez killings. Somehow, the trail of Ed Gein and Terry Caspersen had led there. The crime itself, with its brutalization of innocents, brought up memories of the deaths of Michael and Alex Smith in South Carolina.

The flat color images that appeared on the computer monitor rendered the blood trail through the Lopez house in precise but somehow unreal detail. The shot of Erik Lopez's tiny handprint had an immediate, terrible impact, but the other shots blurred together. I found myself analyzing them in terms of blood-spatter patterns. The blow had to come from here, the skull positioned there. Order from disorder.

An often-repeated fact about television viewing is that by the time an average American child leaves grade school, he or she will have witnessed eight thousand murders. In present-day media culture, some of these statistics become unmoored from their sources and take on a life of their own (like the idea that to remain healthy one must drink eight glasses of water a day). But the "eight thousand murders" stat has impeccable credentials, first published in a 1992 report from the American Psychological Association, detailing a content analysis study conducted at the

University of Pennsylvania's Annenberg School for Communication by George Gerbner and his associates.

Gerbner was dean of the Annenberg School and a leading proponent of Cultivation Theory, including a subthesis he developed, the "mean/scary world" theory. The latter states that people who are exposed to a lot of violence (Gerbner was talking about television, but the same could be applied to Aftermath techs, police detectives, or forensic scientists) tend to view the world as more menacing than it really is. Asked what the chances were that they would be involved in violence in any given week, heavy television viewers estimated one chance in ten, when the statistical truth was less than one chance in a hundred. In other words, their worldview was skewed by violence on TV.

Before we were married, when we were headed into the ceremony, the Moral Compass gifted me with a tiny scroll, on which was written a quote from Gertrude Stein: "Considering how frightening everything is, it is comforting to know that not much is really dangerous."*

Which comes first, the murder or the fear? Hamlet would have counted himself "a king of infinite space," were it not for the fact that he, like Greg Banach, was troubled with bad dreams. Most of the time, I slept like a baby after jobs, untroubled by dreams lousy or otherwise. And I was usually fine during the job itself, too busy working to be anything except a little dyspeptic.

But in a mental process familiar to anyone who has studied post-traumatic stress, my experiences on the job came back to haunt me. Every once in a while, at random times day or night, my mind would seize up like a rusty chainsaw, fixating on an im-

---

* We later discovered that my wife got the quote wrong, and what Stein really said was, "Considering how dangerous everything is, nothing is really very frightening." I like the first way better.

age from my murder biography that busted unbidden into my thoughts—blood in a basement room filled with fish tanks, say, a cold southern lake, a nine-year-old child's bloody handprint, or a girl on a riverbank in a gingham dress.

"Look there, brother baby," ran the riff in Robert Johnson's "Sweet Home Chicago," "and you'll see what I have seen."

CHAPTER TWELVE

Cold

CSI graffiti

Come lovely and soothing death, undulate round the world. —Walt Whitman

Hey, Mr. Tambourine Man, play a song for me.
—Bob Dylan

Aftermath threw its Christmas party at the Foundry, a sports-bar restaurant on the perimeter of the Westfield Fox Valley Mall in Naperville. It was a somewhat dispirited affair, with a steam-table buffet and the Seahawks-Eagles *Monday Night Football* game on three big-screen TVs that braced the party space. Because of the ebb and flow of Aftermath business, which saw a dip every year in the dim winter months, some of the techs hadn't been working for a while.

Everyone was on their best behavior. The spouses who came with their husbands sat primly beside them. Since the techs usually spent so much time in the field, apart from the office personnel, the groupings at the party became quickly Balkanized. Techs sat with techs, office workers with office workers. Chris handed out cigars and Nancy Doggett, the office supervisor, dispensed the envelopes containing the staff's Christmas bonuses. Given the paucity of jobs during that period, holiday cheer balanced with a surly mood of economic desperation, and Greg and Greg didn't bother to show up.

I left the table and headed for the smoking area around the pool tables. Away from the ears of Chris and Tim, the techs bitched about the lack of work and the tough nature of their jobs. They reminded me of the old joke, two *alter kockers* complaining about their Catskills resort: "The food here is terrible!" "Yeah, and there's not much of it!"

Out of earshot of the techs, Tim bitched, too, mostly about the pressures that the exigencies of the marketplace put on them.

"It's like a vicious circle," Tim said. "We pay a premium on workman's comp insurance. When we can find an insurance company who will write us, they keep jacking up the rates. Then we have to pass those charges along to the people who pay for the jobs. And guess who they are? Insurance companies. We don't make money. We recycle it."

Chris came by with a cigar, sunny and smiling. He looked like the only happy one at the whole party. Then Ryan introduced his girlfriend, a property manager who worked with his mom. Dave's girlfriend, a hairstylist, couldn't make it, but the affair took on an all-in-the-family atmosphere.

I gravitated to Joe and Kyle, the third crew. Joe's wife, Holly, was visibly pregnant with their second child. Ryan congratulated him. "That's additional proof of your manhood," he said, laughing.

Joe held up his Long Island Iced Tea and toasted absent friends. Where were Banach and Sundberg?

"Greg Banach is the best remediation tech in the company," I said rashly. "Which means he's probably the best remediation tech in the world."

I could tell no one agreed with me, but I blundered on. "If his head doesn't explode, he'll stay the best."

We talked about movies.

"Hollywood, you are one sick, sick puppy," Joe assured me. A movie that I had a screenplay credit on, called *Dirty,* had just come out, and DVD copies made the rounds of the company. "That movie was one twisted motherfucker."

You know you have done your job when your stuff grosses out an Aftermath tech.

I had flown in simply to go to the holiday party, since in the course of working alongside the techs I numbered so many of them as my friends. That evening the whole Aftermath enterprise struck me as impossibly fragile. One insurance rate hike, the techs quitting to take other, saner jobs—I didn't know what I thought might happen, but I imagined the company under assault. Catastrophic thinking was the Christmas bonus that Aftermath had given me. In the garish light of the sports bar, faces shone with a green, sulfhemoglobin tint. Or maybe that was just the result of Chris's cigars.

Then something impossible happened. A job came in, one that would eventually help crystallize the whole Aftermath experience for me. Tim took me aside and told me about a homicide cleanup in Evanston the next morning. Greg and Greg were going to handle it, but since they had skipped the party, Tim decided to give it to the second crew, Dave and Ryan.

"My flight's not until afternoon," I said. "I could go to the job, then shoot out to Midway." I felt elated, then immediately felt guilty for feeling elated over someone else's tragedy, then felt elated some more because I had so clearly and so thoroughly entered into the topsy-turvy reality of Aftermath.

"A single mother," Tim said. "They don't know who did it."

On a frosty December morning I followed Dave and Ryan's Aftermath truck as it nosed through a dense residential neighborhood off Dodge Avenue in Evanston. We had difficulty with our approach to Ashland Avenue, a two-block-long street that seemed locked within a maze of one-way thoroughfares. We spotted the building because of a small shrine set up out front, flowers and guttered candles. "Linda we love and miss you," read one message.

The shrine occupied the concrete front stoop of 1144 Ashland, a tan-colored two-story brick apartment house crowded uneasily among single-family residences. Linda Twyman, a forty-three-year-old travel agent and divorced single mother, had occupied the back apartment on the right-hand, northern side of the building. We geared up and entered a living space that clearly had been heavily forensically investigated, with signs of police presence everywhere. Beneath the dusting of several different types of fingerprint powder, though, blood spatter showed up clearly.

The apartment's front entrance led into the living room. Straight ahead, a hallway gave access to two bedroom doors and linen closet before it ended at the bathroom. Through the living room was the kitchen, where the back door led to a side yard.

Fingerprint dust, leuco crystal violet, covered door frames, light switches, doorknobs, doors, window frames, and walls, with some overspray on the carpets. Detectives had removed and taken with them the doorknobs from the linen closet and the first bedroom. Some black ninhydrin dust showed up against the mottled purple of the leuco crystal. Blood stained the gray carpet in the hall and both bedrooms, with some splatter in the living room also. Blood droplets spilled over a clutch of family photos on a low table under windows on the north side of the living room.

Investigators had also sawed out and taken along a section of doorway molding between the kitchen and living room. They lifted linoleum floor tiles from the kitchen itself, which was heavily dusted. In the first bedroom, evidence of the coup de grace. A platter-size black-red bloodstain dried on the floor. On the interior wall, detectives marked penciled lines through the blood splatter, each line lettered *A* through *F* and tracing the trajectory of an individual droplet of blood splatter. The vector lines converged at a point about forty inches off the ground.

Evanston police detectives, who got help from the North Regional Major Crimes Task Force, left behind small one-by-two-inch stickers everywhere. Some were evidence markers, lettered or numbered peel-and-stick decals with calibrations along two sides, utilized to mark areas for crime scene photography. Others gave warnings to those who would come after them. "Biohazard," read one Day-Glo orange sticker. "Possibly Biologically Contaminated. Handle with Gloves." Others, similarly brightly colored, marked off chemically treated areas, or blared out, "Warning Warning Warning BIOHAZARD Blood Blood Blood."

"At least they marked out the places where there are chemicals," Ryan said. "They don't usually do that."

Blood soaked the mattress in the bedroom, and we wondered how we would maneuver it out of the apartment. I followed Ryan through the kitchen and out the back door into the yard. Here, more signs of the police investigation. The police had carefully raked the backyard, clearing away a scrim of leaves and snow.

"Looking for the weapon," Ryan said. He decided to take the mattress out on the opposite side of the building, through the window off the hall, and into the side walkway to the street.

When we returned to the kitchen, Ryan plinked his finger against the bowl of a solitary goldfish. "Saw the whole thing," he said.

"What do you think?" I asked him.

"Oh, this was a hit," he said with utter certainty. "They knew who they wanted, and they came looking for her."

"They?"

"The upstairs neighbor heard screams, looked out her back window, and saw two guys in hoodies running away."

I wasn't convinced. "She was a single mother, gainfully employed at a travel agency," I said. "Who would want to kill her?"

"You never know what people are into. Only the goldfish knows." He brought his face close to the bowl. "And you're not talking, are you?"

Maybe it was the timing of the killing, during the holiday period between Thanksgiving and Christmas, or perhaps it was jet lag, but the contrast distressed me: the heavy forensic treatment of the detectives overlaying the tenuous, gentle existence of Linda Twyman on display in the apartment. Cold rationalism versus the complex warmth of day-to-day life.

In the living room a stack of meditation (*Ocean of Love*) and stress reduction CDs had toppled over onto the floor. The DVD for *Pay It Forward*, Haley Joel Osment's paean to goodwill, lodged in the player. Twyman had decorated the wall above her couch with a simple inspirational poster: "The poor long for riches, the rich long for heaven, but the wise long for tranquility."

"The evidence speaks to me," says Gil Grissom. But there was nothing that fingerprint dust could say to the reality of Linda Twyman's freezer, for example, jammed full of a dizzying array of ice cream treats from Klondike whole fruit bars to Edy's Gourmet butter brickle. The cops investigate the crime. They miss the person.

And yet, at the same time, the clinical procedures of forensic detectives can be oddly comforting. In the face of an event about which nothing can be done, they represent something we can do. They present an objective view of the scene, and refrain from getting caught up in the clutter of subjective judgments, such as what it meant that a single woman found it necessary to reward herself with frozen confections. Overpowered by the emotional weight of Linda Twyman's existence, I took refuge in crime scene science.

Gil Grissom uses a fluorescent fingerprint dust of his own custom mix, which he calls "Red Creeper." Ninhydrin, or Triketohydrindane hydrate, used by detectives in the Twyman case, reacts with amino acids left behind in fingerprints. Investigators can mix it with either acetone or another solvent, such as methanol, propanol, or petroleum ether. It is a toxic, headache-inducing substance, which is why CSIs thoughtfully left stickers behind announcing its use.

The presence of leuco crystal violet (LCV) implied murder, or at least violence, since it was used to raise fingerprints made in blood. In the language of forensics, blood was the "transfer medium" of the prints in question, and it catalyzed the LCV crystals in reaction to hydrogen peroxide. A coloring reagent used in photography and printing, LCV turns bright violet in sunlight and was used generously by CSIs processing the Twyman apartment. The hallway looked as though Prince had been through, painting the walls purple.

Forensic investigators also used more traditional fingerprint powders, made from either charcoal or aluminum dust, which Aftermath techs hated, since they were fiendishly difficult to remove from surfaces. The worst method, as far as the techs were concerned, was Super Glue fuming, the use of cyanoacrylate ester treated with sodium hydroxide and mixed with fluorescent dye. Special Super Glue wands took the process out of the laboratory fuming cabinets into the field. A treated surface is sticky and blackened, resembling the marks left behind by peeled-off price stickers.

"I don't even try with that shit," Banach told me. "No way you can clean it off. If it's on a door frame, I just strip the whole piece of molding right off the wall."

Scientists enjoy using five-dollar words for fingerprint technology, such as *dactyloscopy* and *ridgeology*, but it all works off an evolutionary quirk that humans inherited from their tree-clinging ancestors: friction ridges on the volar surfaces of the hands and feet. Primates have fingerprints for the same reason tires have treads.

In the Twyman case, the fingerprint dusting yielded dozens of prints. Detectives would have to eliminate "knowns" or "residents" prints from the occupants of the apartment. Then they would have to match the remaining, nonresident prints against national computer databases such as the FBI's IAFIS. (The two major federal government databases, IAFIS and USVisit, generate an average of seventy thousand positive ID hits every day.) Having identified suspects on the basis of a match, police would still face the task of locating them, questioning them, building a case against them.

Likewise, blood spatter pattern (BSP) evidence has inherent limitations. Only in rare cases can it lead to identification of the perpetrator. A known case involved a killer who, from the BSP at the scene, had used his right fist to beat the victim, and cut his own hand while doing so. Detectives located a suspect with a fractured right hand that sported a nasty cut, who eventually confessed to the crime.

Generally, though, BSP evidence points not so much to "who" as to "how." After long study, a base of knowledge has grown up to characterize the behavior of blood droplets under various conditions. To solve for the angle (A) from which a blood droplet fell, for example, CSIs use a simple formula:

$$A = arc \sin \left(\frac{width}{length} \right)$$

Blood spatter demonstrates a signature teardrop shape that indicates direction of travel: from the "fat," rounded end toward the thin, pointed end. Generally, the smaller the droplet, the greater the speed of travel, from the mistlike spatter from gunshots to the round "crown" droplets from stationary sources. Drops thrown off from the bloody hair of a human in flight array themselves in bizarre, complex patterns.

Reading the blood spatter in Linda Twyman's apartment did not reveal a pretty scenario. Twyman was first wounded in her kitchen, ran into the living room (where she bled over her family photos) and was caught again in the hallway. She stumbled into the bedroom off the hall, smearing the doorjamb in the process.

In the bedroom, the assailant delivered the blow that killed her as she knelt on the floor. The convergence of the pencil lines on the interior wall established that the blow came forty inches off the floor, which was consistent to the kneeling height of a woman Twyman's size. A cast-off pattern—blood flicked from the blade as the killer drove it downward—indicated the force of blow. The massive bleed-out stain marked the place of her death.

Every contact leaves a trace. Locard's exchange principle summons up a universe where every action imprints itself indelibly upon the world, which we could read if only we had an apparatus of observation sensitive enough to do so. We lift our hand and displace atoms of air, and those atoms could be read and tabulated. A sparrow falls, and that could be marked also.

BSP analysis traced Linda Twyman's last moments in the world. But as she moved through her life she touched many people, her family and friends, in ways not examined in any forensic textbook. As I stood in Twyman's apartment, all the multiple signs of a thoroughgoing forensic investigation appeared paltry to me. I could see clearly that the effort to understand her death failed utterly to illuminate her life. "We don't know a millionth of one percent about anything," said Edison.

Imagine we trail behind us shimmering life lines that trace our progress through the world. They mark our everyday trips to the grocery store, back and forth to our places of employment, far afield to our Maine vacations. Our residences, naturally, would be scribbled thick with such lines, drawn and redrawn repeatedly as we move back and forth to the kitchen, bathroom and bedroom.

Now mark the lines of a murderer. Mark the lines of a victim.

In the great majority of cases, because most victims know their killers, the life lines of murderer and victim demonstrate a veritable crosshatch of interrelatedness. The two paths intersect many, many times before that of the victim suddenly terminates. But in rare cases, the life lines of killer and killed pursue their meanderings quite independently, to cross once and only once, at the moment of murder.

These are the murders we fear most. Charlie Manson's group of hippie killers got lost in the L.A. canyons, knocked on the wrong door, and wound up barging in on a pregnant movie star. The life lines of Sharon Tate and the Manson girls had never crossed before. These kinds of killings seem terrifyingly random. But fear of them is irrational. Statistics tell us we should be looking at our family, our relatives, our friends and neighbors. But statistics don't lessen the terror.

In Linda Twyman's case, the principle of "look closest first" yielded no suspects. Her boyfriend of four years was ensconced at a gated army base three hours to the north, with entry and exit closely monitored. Her ex-husband was across the country in Minneapolis with the couple's daughter. The life lines that crossed and recrossed led nowhere. There was left the chilling possibility of a random, wrong-place-at-the-wrong-time murder.

The killing of Linda Twyman enjoyed only a very brief run in the local newspapers and on television stations. The death of a single mother was not particularly good copy. Evanston police, playing it very tight-lipped, were no help. Although it had help from the North Regional Major Crimes Task Force, an ad hoc expertise-sharing investigative body, the Evanston police department allowed the trail to grow cold. Linda Twyman's murder appeared as a blip on the radar, glowed momentarily, and then winked out.

....

In February 1964 a twenty-three-year-old Bob Dylan journeyed in a new Ford station wagon he had just bought, through the South to New Orleans, where he was enthralled by the spectacle of a jazz funeral procession. A frequent aspect of such pageants was a solitary mourner, carrying a tambourine, walking ahead of the hearse. Dylan subsequently wrote "Mr. Tambourine Man," a song I listened to for much of my life without realizing the lyrics were an invitation to death.

Whitman, too, in his epic elegy for Abraham Lincoln, "When Lilacs Last in the Dooryard Bloom'd," wrote "a chant of fullest welcome" to death. This clashes so sharply with the usual human stance of abhorrence and avoidance that we must leave it to the poets to accomplish it. But in spite of the famous directive from another Dylan—"Do not go gentle into that good night"— some level of reconciliation with death could be said to be the goal of most religious and philosophical thought.

Their experience at the helm of Aftermath has left Chris and Tim with diametrically opposed ideas on what Chris calls the "big questions." He and Kelli described themselves to me as "definite Christians" who attend church every week and are committed to being part of the community of believers.

"I believe this life is a test," Chris said. "How well you act on earth, how well you behave—do you walk with the Lord?—those are all questions that are going to be decided in the here and now. But I also think the here and now is just like the blink of an eye. You are here and then you're gone."

Chris responds to the constant exposure to tragedy with something of a "seize the day" philosophy. "A lot of times the deceased wasn't doing anything wrong, anything that would put them in harm's way," Chris said. "All of a sudden, someone comes upon them and stabs them, or they get stuffed through a machine at work. It doesn't make any kind of sense that you can understand. So you've got to be ready to go at any moment."

Tim Reifsteck went the other way. He, too, discerns a lack of understandable pattern in the stories of death and dying he encounters on the job. "I was asking myself a lot of questions back when we started Aftermath, and even before," he said. "I put a lot of thought into it. At a funeral, the pastor says, when you lose a loved one, 'Well, God needed him.' And I think, what a selfish thing to say to me! That someone needed this loved one more than their family or their kids? I cannot believe that. I cannot rationalize that."

Tim's wife, Sara, attends a small, nondenominational Christian church, and is raising their two sons as Christians. "It's been like baby steps," she said. "I call myself a baby Christian. I feel like God's watching me more since I had kids."

She became a churchgoer after watching one of her friends suffer from uterine cancer at a young age. "The disease turned her inside out," Sara said. "She looked like a skeleton by the time she died."

The same experience would have spun her husband off in a totally different direction. "So there's some higher being that looks down on this earth and says, 'You know what, I am going to take a mixture of myself and put it in people and see how they grow up'? I can't accept that at all. If it were true, why would he allow children to be raped and murdered, women to be raped and murdered? Because that's happening every second of every day."

Linda Twyman wasn't raped, but her murder challenged notions of the good death—a death that comes naturally in the course of a long and rewarding life. Chris and Tim respond to that challenge in different ways, Chris by becoming more deeply religious, Tim by rejecting transcendent meaning as a guiding principle of life. But from their experiences at Aftermath, both of them understand the brutally random nature of death.

"I was in downtown Chicago, and I'm looking at this guy standing at the curb," Tim recalled. "Actually, he was standing with his feet hanging over the curb. And I thought, 'Buddy, if you only knew how close you were to dying.' Because maybe a little old lady takes a corner too sharp—oops!—and that's it."

In the Aftermath universe, random deaths qualify as bad ways to go, along with deaths by violence, unattended deaths, and those marked by dementia. When I asked Chris and Tim to describe the way they imagined their own deaths, they both answered the same way. They wanted to be surrounded by people they love.

"The good death," a phrase that has been around since Chaucer's time, embodies this concept in the West, but the idea itself is ancient as language. It has, at various times, been interpreted in different ways. In warrior cultures, the good death was the heroic death on the battlefield, with the Valkyries ushering the dead hero to immortality in the halls of Valhalla. In Chaucer's day the Ars Moriendi ("art of dying") movement held sway in Europe, arguing that the way to live was to prepare for death every single day you were alive.

Ars Moriendi had a fetishistic aspect called memento mori. The phrase is Roman in origin, and means "remember death," but it was the Middle Ages that really took the concept to heart. The idea of keeping the thought of death always in mind yielded up some of the strangest art known to man. "Cadaver tombs" featured sculpted renderings of decomposing corpses: a representation on the outside of the tomb of what was going on inside. Monks went goth, keeping skulls around to contemplate. Paintings of skeletons, petal-shedding flowers, or snuffed candles embodied the memento mori theme in art.

Today we are in the grips of *obliviscere mori*, "forget death." Death is carefully walled off from our day-to-day existences, behind words like *morbid* and *gloomy*. A cheerfulness prevails that can appear a shade desperate. Up with people, down with death. We hire firms such as Aftermath to deal with it. If we think about it at all, the good death is pretty universally regarded as a painless one, or passing away in sleep with loved ones at the bedside. For some people, the good death represents an oxymoron. No death is good.

Linda Twyman could not be said to have died the good death. She died alone, killed by strangers with knives who stabbed her repeatedly about the head and neck, with the coup de grace coming from a cut that severed her jugular. Medical examiners classify deaths in different ways. They distinguish among the manner, cause, and mechanism of death (some MEs throw in mode of death too). In Twyman's case, the manner of death was ruled homicide, while the cause was an incise wound and the mechanism exsanguination.

Medical examiners sometimes call an autopsy a "narrative," because the procedure teases out the story of a particular death. My own narrative of Linda Twyman's murder—a cheerful and life-affirming forty-three-year-old woman living in a run-down neighborhood of an otherwise wealthy municipality, attacked by person or persons unknown, her killing left unsolved by police and forgotten by all but those who knew her—affected me more deeply than any other death I encountered at Aftermath.

There is the physical crime scene, the number two pencils inserted in bulletholes and the walls smeared with aluminum fingerprint dust. But there is an emotional crime scene, too, of affected friends, relatives, neighbors, bystanders. That scene could not be cleaned up quite so tidily. The emotional crime scene is much more long lasting. What the Aftermath techs did might have an impact on it, in the sense that the physical always affects the emotional. But Ryan or Dave could never clean it up entirely. Only time would do that.

Death of a loved one invests the survivors with a fierce claim to privacy. The egotism of loss is absolute. Society assumes grief to be an intensely personal, privileged activity. Our sexual lives are no longer secret, but our mourning periods remain sacrosanct. No trespassing allowed.

Well, I trespassed. I violated. I hoped I did it in the manner of Gunther von Hagens, say, rather than H. H. Holmes, for positive reasons rather than selfish ones.

When a Topeka, Kansas, hate group led by Reverend Fred Phelps started showing up at military funerals, to protest loudly with signs and slogans that war deaths in Iraq and Afghanistan were God's punishment for the country's wicked, decadent ways, state legislatures quickly pushed through laws to protect the sanctity of the rites.

Death be not fucked with. I once went to a ceremony in a New Jersey cemetery, where the four grandsons of the deceased beat up a cabbie and put him in the hospital, because he had honked his horn while their grandfather's casket was lowered. My father, an unreconstructed gay-hater, physically attacked his homosexual cousin at his mother's funeral, punching him in the face and throwing him to the ground. The cousin's trespass? Bringing his boyfriend to the service.

One could argue that these physical attacks, not a honking horn or a gay boyfriend, were the real violation of the atmosphere of dignity and decorum that ought to be maintained. Grief grants a sense of self-righteousness that can bring out the worst in people, setting them firmly on their high horses.

A physical crime scene, an emotional crime scene, and perhaps a spiritual crime scene too. Some biblical scholars contend "Thou shalt not kill" is actually a mistranslation. What the Decalogue really says is "Thou shalt not murder," thus letting off the hook warriors, government executioners, and law enforcement officers who kill in the line of duty. Whatever way the meaning is shaded, homicide violates a universal, fundamental, and age-old law, with biblical if not spiritual consequences.

When they were married in the 1980s, Linda Twyman changed her then-husband James Twyman's life by giving him a book, *Autobiography of a Yogi*. The account by author Paramahansa Yogananda became something of a viral phenomenon after its publication in 1946, spreading Vedic philosophy and the practice of yoga through Yogananda's adopted home in the West. The New Age movement claimed *Autobiography* as one of its seminal texts, and the book has influenced luminaries as diverse as Mahatma

Gandhi (who was portrayed in its pages), Elvis Presley, and Jack "Chicken Soup for the Soul" Canfield.

It did a number on James F. Twyman. He embarked upon a twenty-year spiritual quest that led him to found the Beloved Community and the Emissary of Light, groups that express a hopeful, peace-through-love New Age philosophy. Under the banner of the "Peace Troubadour," Twyman began traveling to the world's war zones to put on concerts and prayer vigils. Armed only with a guitar, he showed up in Croatia, Sri Lanka, Northern Ireland, and the West Bank, hosting gatherings that over the years numbered in the hundred of thousands.

Linda and he divorced but remained close. James became one of the leading lights of the New Age scene, authoring such books as *The Art of Spiritual Peacemaking*, *The Secret of the Beloved Disciple*, *Portrait of the Master*, and *Emissary of Love*.

"While I traveled to some of the most dangerous places in the world, Linda was murdered at home, in her own apartment," Twyman wrote in an e-mail.

Twyman devotes all his time and energy to plumbing Chris Wilson's "big questions." He has evolved a death-is-not-death philosophy and shares a lot of ground with his friend and spiritual ally, Neale Donald Walsch, author of the best-selling Conversations with God series.

"Linda's death has taught us so much about life, and about the unreality of death itself," Twyman wrote. "We now have another angel at our side creating peace in this world."

I didn't sense any angels in Twyman's apartment when Ryan and Dave addressed the mess the police left behind a week after she died. But I hadn't devoted my life to spirituality, so perhaps I was just blind to them. The equanimity and grace demonstrated by James Twyman in the face of his ex-wife's death indicated a path, arduous and long, toward Whitman's "chant of fullest welcome."

"You're not going to turn all softheaded on me, are you?" the Moral Compass asked during a late-night phone call. In my Extended Stay purgatory, I had come to depend more and more on talking to my family to keep the black dog of depression firmly on its leash.

"I thought you loved Walt Whitman," I said.

"Of course I do," the Moral Compass said. "But that doesn't mean he can't be softheaded sometimes."

"You can hardly blame me for thinking about this stuff," I said. "I'm just trying to figure out how it fits into the grand scheme of things."

"There is no grand scheme of things," my wife said. "Unfortunately."

Hurricane Katrina had kicked the stuffing out of the South just three months previous, and we talked about the devastation and dead bodies that were still turning up down there. Aftermath had made a corporate decision not to get involved in the relief effort, judging that the field was already overcrowded. Chris and Tim sent equipment donations instead. I quoted my wife a line from a raucously mournful blues song of ZZ Top: "Jesus just left Chicago and he's bound for New Orleans."

The Moral Compass sighed. "You don't need to go to New Orleans," she said. "Just come home safely."

For me, the presence of all the fingerprint dusting, police stickers, and other investigatory leftovers in Linda Twyman's apartment served a contradictory purpose. Instead of grounding what had happened in the just-the-facts-ma'am world of forensics, the police effort made me want to locate Twyman as a human being. She was not just "bio," as the son-in-law of another dead man put it. And her apartment was not just a scene to be processed by all the advanced techniques forensic science had to offer. It had been her home.

Even more than forensic detectives, Aftermath deals in the most rudimentary, physical aspect of existence. But Aftermath allowed me a window into the most spiritual aspects of life too. As Upton Sinclair wrote, knee-deep in gore in the Union Stock Yards, "One could not stand and watch very long without becoming philosophical." Over the course of my time at Aftermath, I had become philosophical with a vengeance: about death and dying, about the human stain, about my own complicity as a voyeur. The Fellow in the Bright Nightgown and I had become, if not friends, then at least much better acquainted.

I left Ryan and Dave still scrubbing the leuco violet off the hallway walls. I closed the front door to the Twyman apartment and stepped out into the insipid sunshine of a midwestern winter afternoon. I had done the last job I would ever do with Aftermath. I drove to Midway Airport, dropped off my rental car, and took a plane for home.

The Dead
House

A ZAKA volunteer at the scene of a bombing

You need any help with the coffin, call me.
—Erich von Stroheim, *Sunset Boulevard*

I've never heard of a crime I could not
commit.—Goethe

The New York City firemen on "the pile," as the hulking wreckage of the World Trade Center was called during its removal process, objected strenuously to the term "cleanup." Its use, they considered, implied that the bodies of their comrades who had died in the attack were so much dirt or refuse to be expunged, scrubbed away, and carted off. They also objected to the location of the processing station for sorting through the debris from Ground Zero, at a former city dump on Staten Island called Fresh Kills. The name was Dutch, *kills* being the Dutch word for *stream*, but again, the fire personnel protested their dead being associated with garbage.

The removal of the dead in the Trade Center attack represented a bioremediation effort on a scale never before attempted or imagined. Aftermath writ large. Of the 2,749 victims who died in the collapse of the towers, only half have been identified, in spite of a massive DNA typing project. The rest of the victims were pulverized by the collapse of a 1.5-million-ton structure, or immolated by fires that reached fifteen hundred degrees Fahrenheit and did not burn out until January 2002, five months after the 9/11 attack. Workers in the recovery effort were reduced to sniffing shovelfuls of dirt, vigilant for any scent of decay.

As part of my work for *Maxim*, I attended press briefings at the New York medical examiner's facilities that handled the bodies recovered from the Trade Center. The facility was an ad hoc affair, some of it spread out beneath canvas tents. Workers at the site referred to it as "the Dead House." While at the Dead House I became interested in the work of a group of Israeli volunteers known by the acronym of ZAKA.

When I told a Jewish friend about working with Aftermath, he said, "My people have been doing that for a long time." He was probably thinking of ZAKA, which stands for the Hebrew words *Zihuy Korbanot Asson,* which in English means "identifying victims of disaster." I didn't realize it at the time, but the ZAKA volunteer I spoke with that day in the Dead House, who gave his name only as Yehudi, was the first bioremediation tech I ever met.

Yehudi was a slight Yeshiva student with a wispy beard and glasses. He seemed nervous among the bustle of the New York media, and uncomfortable talking to me, who was clearly a goy and far from his ultra-Orthodox roots. He carried with him but did not wear the ZAKA vest he normally donned whenever responding to emergencies. The fluorescent "safety yellow" color of the vests rendered the volunteers instantly recognizable at disaster scenes.

"I started doing this work two years before," he said in accented English. "Now I am here in New York to help." He told me the Torah declared that working to protect the sanctity of the dead was among the highest mitzvahs, or good deeds.

"The dead cannot repay the favor," Yehudi explained. "So the mitzvah is pure and without motive."

The strictures of Orthodox Jewish law are exact about the treatment and purification of the dead. Whenever possible, burial takes place within twenty-four hours. Bodies of the dead must be protected from desecration. An Orthodox burial society, or *hevra kaddisha,* will even provide watchers to guard the body. Most important, however, is the need for the body to be buried as intact as possible. This means that the gathering up of every particle of biomatter takes on a religious significance.

ZAKA was founded in 1989, and grew out of the horrendous task of applying the principles of Jewish law to the recovery of Israeli bombing victims. For the volunteers of ZAKA, there was no other option: Every shred of flesh, fingernail, skin, tooth, or bone had to be religiously reclaimed for burial. ZAKA developed into a first-response organization that worked not just on the Trade Center attack, but the *Challenger* disaster and the Asian tsunami relief efforts.

Remediation comes from Latin roots meaning to "heal again." At its highest level, at the ZAKA level, the kind of work Aftermath does can reach a healing, spiritual plane. Aftermath is unabashedly a commercial enterprise, but in the unlikely sensitivity of its techs—most of them caustic young ex-jocks, and the last people you might expect to embrace the spiritual aspect of anything—I detected the kind of deep, age-old respect for the dead that ZAKA runs on, and which so many health professionals and hospice workers participate in too.

After working alongside the techs over the course of a year at several dozen Aftermath jobs, I was still unsure where I fit into the equation. I had developed a tremendous respect for the work they did, especially their sure-handed manner of dealing with emotionally fragile people. They seemed half businesslike and half empathetic, an odd combination that should not have worked, but did. Recognizing that I didn't have the innate sensitivity demonstrated so effortlessly by Ryan and Dave, and even by Greg Banach, I rarely spoke to bereaved clients for fear of sticking my foot in my mouth.

But I also recognized that my attraction to Aftermath work was not, strictly speaking, kosher. I was a spy in the house of the dead, presenting myself under false pretenses. Was I a tech or a voyeur? Was what I was doing a mitzvah or a betrayal? And I felt the same conflict more broadly within myself. I was repelled by crime, violence, death, what Arlo Guthrie called "blood and guts and veins in the teeth." I was also drawn to it.

A spy in the house of the dead. The background of death and violence at Aftermath job sites could be gruesomely spectacular, but another aspect of the work, the ordinary evidence of lives lived, affected me almost as much. I felt guiltily privileged to enter the personal spaces of the deceased. The mundane, everyday textures of American life spilled open to me in the bedrooms, kitchens, and upstairs-over-the-garage spaces where Aftermath worked. The clutter of the households, frozen at the time of death, was somehow transformed by tragedy into something engrossing, meaningful.

My father used to call the obituaries in our local newspaper "the meaning pages." In the rooms I entered as an Aftermath crew member, I read an obituary of objects. Congeries of old mail, a yellow bowling shirt with a red-ink Bic pen in its pocket, a stack of *National Geographic* videos, a book on real estate left open spine-up, a refrigerator magnet snapshot of a granddaughter, a small Tupperware tub filled with pennies, a collection of empty amber-plastic prescription containers—on and on it went, the detritus of interrupted personhood, varves of biographical sediment no longer accumulating.

This, as much as the coagulated blood and embarrassing spill of body fluids, left a deep impression on me. The commonness was wrenching. There was the familiar chalk-outline quality to it. What we leave behind defines us. It made me want to run home and pick up my house, toss a whole lot of things out.

"Master, what is the meaning of the universe?" said the young student in the Zen koan, to which the Master replied, "Clean your rice bowl."

An aura of moral certainty attracts both writers and readers to crime. Crime fiction, at least, is a world of white hats and black hats. Even when portraying ambivalent heroes such as Hammett's Continental Op or Chandler's Marlowe, the author usually manages to come down squarely on the side of what's right. For the weak, against the strong. In less capable hands, this Manichean crime-thriller world degenerates into a rigid catechism of snarling avengers battling evildoing monsters.

In the most expansive view of the subject, a great deal of world literature can be classified as true crime, from *Gilgamesh* and *Hamlet* to *The Passion of the Christ*. But for me, the world of true crime, like the world of Aftermath, was the opposite of clear-cut, morally certain territory. It was a deeply compromised place, just as messy and difficult to clean up as a job site itself. Yes, I knew and honored the traditional forms of respect for the dead. But neither could I shake the relish, the satisfying sense of not-me, the voyeuristic thrill.

I couldn't quite adopt only the pure white hat, as Ryan, Dave, and the other techs did so effortlessly. I was somewhere between the white and black. I wore herringbone. I identified with the dead, to be sure, but I also identified with the murderers. In the right (or wrong) circumstances, I could all too easily imagine myself with a knife in my hand.

"If my thought-dreams could be seen," sang Dylan, "they'd probably put my head in a guillotine."

I emerged from the bloody cocoon of Aftermath not quite as a butterfly, more like a mangled moth, a night creature still attracted to the flame. I recall easing back into my "normal" life in quiet, crime-free Westchester. I was restless. I couldn't really share what I had done with my wife and daughter, and it bothered me to be set apart from them. I called the techs repeatedly, wanting to be filled in on all the gritty details.

"Joe said they're really busy," I told my wife after one of these calls. Then I blurted out, "I wish I was out there!"

I was caught in the middle of a conversion experience. People were still dying out in the suburban wastelands of Chicago, but I had returned to my sanitary life. I could leave it all behind. The techs couldn't, and even more to the point their clients could not.

"You're in withdrawal," the Moral Compass said. "You need some sort of twelve-step program."

I felt estranged from my suburban world. I recall one afternoon at the local community pool, describing to some friends what happened to Donald Gene Buchanan on the tarmac out on El Paso. I acted out the crouch Buchanan assumed as he worked on the CFM 56 engine, then the straightening up and the fateful single step forward. I got into it. But looking to my audience, a handful of soccer moms and dads, I suddenly registered their uneasiness and disgust. Young kids packed the concrete apron around the sunlit pool. I was casting shadows. I could have made the point that there was no difference, no separation, that sunlight and shadow always coexist, but I didn't. I finished my story with an abrupt, lame joke, and my audience moved away as from a bad smell.

But the coin had another side. I was at one of the Moral Compass's book parties, when two of-a-certain-age women buttonholed me and asked what I was working on. I remained vague, but they wormed it out of me. Their eyes began to sparkle avidly as I spoke about my Aftermath experiences. They smiled and nodded. They wetted their lips with pink, prehensile tongues. They began to scare me.

"And the blood," one of them asked with obvious savor, "were you actually up to your elbows?"

Eventually, I regained my balance. I still had post-traumatic flashes of blood and guts and gore. Somehow a photo of an Aftermath job wound up stuck in the sun visor of the family car ("Tinley Park suicide" was marked in pen on the back of the photo, "Banach, Sundberg, Bryan"). On the front, a massive, spectacularly variegated pool of blood showed black, brown, vermillion, brick, purple, pink, and yellow against a linoleum background. In the pool's center lay a fat white rubber band. Every once in a while for a few weeks the photograph would drop down onto my lap as I was driving.

But another feeling grew in me, gradually dominating the dread. I realized I was thrilled to have had the wealth of experience that working with Aftermath had granted me. I marveled at it. It became a source not of dismay, nightmare, or estrangement but of secret pride.

During the last years of his life my father increasingly dwelt on a twenty-six-month stretch in his early twenties when he was in Europe with the army air force. He loved talking about his experiences. He fully inhabited the cliché of a veteran garrulously reminiscing over his war years.

"It's only natural," said the Moral Compass, after a bout of listening to my father gas on. "Think about the lives of these men. They were just hometown guys, teachers or mechanics or, like your father, salesmen. Going to Europe and fighting the Nazis? That was the most exciting thing that ever happened to them. By far. Of course they go back to it. How could they not?"

The horror, the horror. Well, maybe the horror drops away, and what I'll be left with is not the foreign, but the human. The techs, the police, the families, the small colony of the saved gathers around the dead, formulating our good-byes, trying to find our balance.

....

Heading into its tenth year of operation, Aftermath continues to expand. Fourteen Aftermath offices cover an area that embraces every state in the lower forty-eight.

Dan Doggett works to enlarge the company's presence in California, employing two of his sons in the business.

Tim Reifsteck's older brother, Kevin, and his younger brother, Bryan, now oversee a vast territory stretching from Ohio south to Georgia and west to Texas.

"In terms of square miles, their territory is about the size of Mexico," Tim said with a laugh. "They've got a lot of marketing to do."

A new custom-built national headquarters for Aftermath, Incorporated, only a few miles away from the old Plainfield office, is under construction.

Chris Wilson sold the Mercedes G55 morgue-mobile back to the dealership from which he'd purchased it. "They called me and said they had another buyer, and would I be interested in selling." He did so, trading laterally for an SL65. Not to be outdone, Tim Reifsteck bought a new F430 Ferrari.

Chris and Tim's labor pool of techs constantly shifts and realigns itself. Greg Banach and Greg Sundberg no longer work at Aftermath. Banach, at least, went out with a bang, literally—slamming the door to the truck bay at the Aftermath offices, jumping it off its hinges after an argument with Chris and Tim. In a more amicable parting, Dave Creager, too, has moved on, joining his brother in a long-distance trucking service. Ryan O'Shea and Joe Halverson now partner as first crew.

The jobs keep coming. Business has never been so good.

Acknowledgments

This book would not have been possible without the generosity and forbearance of Chris Wilson and Tim Reifsteck, founders, owners, and operators of Aftermath, Inc. They invited me into their company, but they also invited me into their lives. Equally necessary was the good faith and good humor of the Aftermath technicians, who put up with a rookie's mistakes and a newbie's nausea, and helped with guidance over difficult terrain. First and foremost were Ryan O'Shea and Dave Creager, who took me into their charge, along with Joe Halverson, Kyle Brown, Dan Doggett, Greg Banach, and Greg Sundberg. Nancy Doggett and Raquel Garcia cheerfully put up with the extra work at the Aftermath offices that my being there entailed.

Erin Moore, my editor at Gotham, combined enthusiasm with clear judgment and an always cheerful mien in the face of challenging subject matter. My agent, Paul Bresnick, and his partnering agent for this project, Byrd Leavell, and Greg Dinkin and Frank Scatoni at Venture Literary helped bring the book to fruition.

The staff at the Westchester County Library System worked tirelessly on my endless requests for material. Greg "Stickman" Klos aided research on the home turf invaluably, while Sally Howe contributed a vital phrase. Josefa Mulaire scanned some of the images. My former editors at *Maxim*, Jim Kaminski, James

Heidenry, and Jason Kersten, taught me a lot about crime writing. Thanks to Lloyd Westerman for making the connection. I would be remiss not to give a shout-out to Sony for their M-630V microcassette tape recorder, technology that is no doubt antiquated in this age of digital chips, but which served me faithfully.

My family provided support, even though they remained mystified about my choice of subject matter. Eloise Reavill recorded her memories about growing up in Chicago thirty blocks to the south of the Union Stock Yards, and about whether the names of Al Capone, Johnny Torrio, and Dion O'Banion were familiar from her childhood (yes, no, no).

Once more into the breach, Jean Zimmerman served as my comrade in arms, the enemy being me making a fool out of myself. The relative lack of success in that battle has nothing to do with the ferocity of her warrior spirit. Maud Reavill provided love, support, humor, and, probably most important of all, distraction. While I was up to my elbows in death, she modeled the life force in an invincible, effortless manner.

Select Bibliography

Baden, Michael, with Judith A. Hennessee. *Unnatural Death: Confessions of a Medical Examiner*. New York: Ivy Books, 1989.

Brenner, John. *Forensic Science Glossary*. Boca Raton, FL: CRC Press, 1999.

Coe, Sue. *Dead Meat*. New York: Four Walls Eight Windows, 1996.

Costello, Peter. *The Real World of Sherlock Holmes*. New York: Carroll & Graf, 1991.

Fisher, Barry A. J. *Techniques of Crime Scene Investigation*. Boca Raton, FL: CRC Press, 2003.

Fletcher, Jaye Slade. *Deadly Thrills*. New York: Onyx, 1995.

Fox, James Alan, and Marianne W. Zawitz. "Homicide Trends in the United States." Washington, D.C.: Bureau of Justice Statistics, U.S. Department of Justice, 2004.

Geberth, Vernon J. *Practical Homicide Investigation: Tactics, Procedures and Forensic Techniques*. Boca Raton, FL: CRC Press, 1996.

Ghiglieri, Michael P. *The Dark Side of Man: Tracing the Origins of Male Violence*. Reading, MA: Perseus Books, 1999.

Grossman, James R. (ed.) *The Encyclopedia of Chicago*. Chicago: University of Chicago Press, 2004.

Holton, Hugh. *Chicago Blues*. New York: Forge Books, 1997.

Iserson, Kenneth. *Death to Dust: What Happens to Dead Bodies*. Tucson, AZ: Galen Press, 2001.

Kobler, John. *Capone: The Life and World of Al Capone*. New York: Da Capo Press, 1971.

Kübler-Ross, Elizabeth. *On Death and Dying*. New York: Scribner, 1997.

Langan, Patrick A., and David P. Farrington. "Crime and Justice in the United States and in England and Wales, 1981–96." Washington, D.C.: Bureau of Justice Statistics, U.S. Department of Justice, 1998.

Larson, Erik. *The Devil in the White City*. New York: Crown, 2003.

Lester, David. *Why People Kill Themselves*. Springfield, Illinois: Charles C. Thomas, 1992.

Lesy, Michael. *The Forbidden Zone*. New York: Farrar, Straus and Giroux, 1987.

Lesy, Michael. *Wisconsin Death Trip*. New York: Pantheon, 1973.

McGinty, Jo Craven. "New York Killers, and Those Killed, by Numbers." *The New York Times*, April 28, 2006.

Maeder, Jay. "Ghost Story: The Collyer Brothers, March–April 1947." New York *Daily News*, September 19, 2000.

Miller, Donald L. *City of the Century: The Epic of Chicago and the Making of America*. New York: Simon & Schuster, 1996.

Nuland, Sherwin B. *How We Die: Reflections on Life's Final Chapter*. New York: Knopf, 1994.

Roach, Mary. *Stiff: The Curious Lives of Human Cadavers*. New York: W. W. Norton & Company, 2004.

Rosso, Gus. *The Outfit: The Role of Chicago's Underworld in the Shaping of Modern America*. New York: Bloomsbury, 2001.

Rule, Ann. *The Stranger Beside Me*. New York: Signet, 2001.

Sachs, Jennifer Snyder. *Corpse*. Cambridge, MA: Perseus Publishing, 2001.

Saferstein, Richard. *Criminalistics*. New York: Simon and Schuster, 1998.

Sante, Luc. *Evidence*. New York: Farrar, Straus and Giroux, 1992.

Sinclair, Upton. *The Jungle (100th Anniversary Edition)*. New York: Penguin Classics, 2006.

Vass, Arpad A. "Beyond the Grave: Understanding Human Decomposition." *Microbiology Today*, November 2001.

ABOUT THE AUTHOR

Gill Reavill has written extensively on true crime and disaster for *Maxim* and other publications. A screenwriter (*Dirty*, a police drama staring Cuba Gooding, Jr.) and collaborator on numerous books, including the *New York Times* bestseller *Beyond All Reason: My Life with Susan Smith*, he has appeared on many television shows, including *Today*. He lives in New York's Westchester County.